FARM AND WORKSHOP
WELDING

Everything You Need to Know to Weld, Cut, and Shape Metal

THIRD REVISED EDITION

ANDREW PEARCE

FOX CHAPEL
PUBLISHING

Frontispiece illustration: *Hardfacing with MIG*
Photographs by Andrew Pearce unless otherwise credited
Book layout by Liz Whatling
North American technical editor Brian T. Yarrison CWE, CWI

Marked photographs on pages 32, 33, and 112 courtesy of ESAB North America.
Additional photo credits: Shutterstock; Back cover, Valmedia, MIG Welding; Salivanchuk Semen, Manual
Metal Arc Welding; Bogdan VASILESCU, Gas Welding; Interior, Aleksandar Malivuk, 4; Valmedia, 5 (MIG
Welding); Extarz, 5 (TIG Welding); Bogdan VASILESCU, 5 Gas Welding and Cutting; Jack Frog, 119.

Projects featured in Bonus Section: DIY Welding Projects (pages 4, 119-134, and Back Cover) are
courtesy of Forney Industries and Hobart Welding Products, as noted. All photos and text are used with
permission.

Farm and Workshop Welding, Third Revised Edition (2021) is a revised edition of *Farm and Workshop
Welding* (2012), published by Fox Chapel Publishing Company, Inc. Revisions include new text, photos,
and bonus section with step by step projects.

ISBN 978-1-4971-0040-4

The Cataloging-in-Publication Data is on file with the Library of Congress.

To learn more about the other great books from Fox Chapel Publishing, or to find a retailer near you, call
toll-free 800-457-9112 or visit us at *www.FoxChapelPublishing.com*.

We are always looking for talented authors. To submit an idea, please send a brief inquiry to
acquisitions@foxchapelpublishing.com.

Printed in China
Eighth printing

Contents

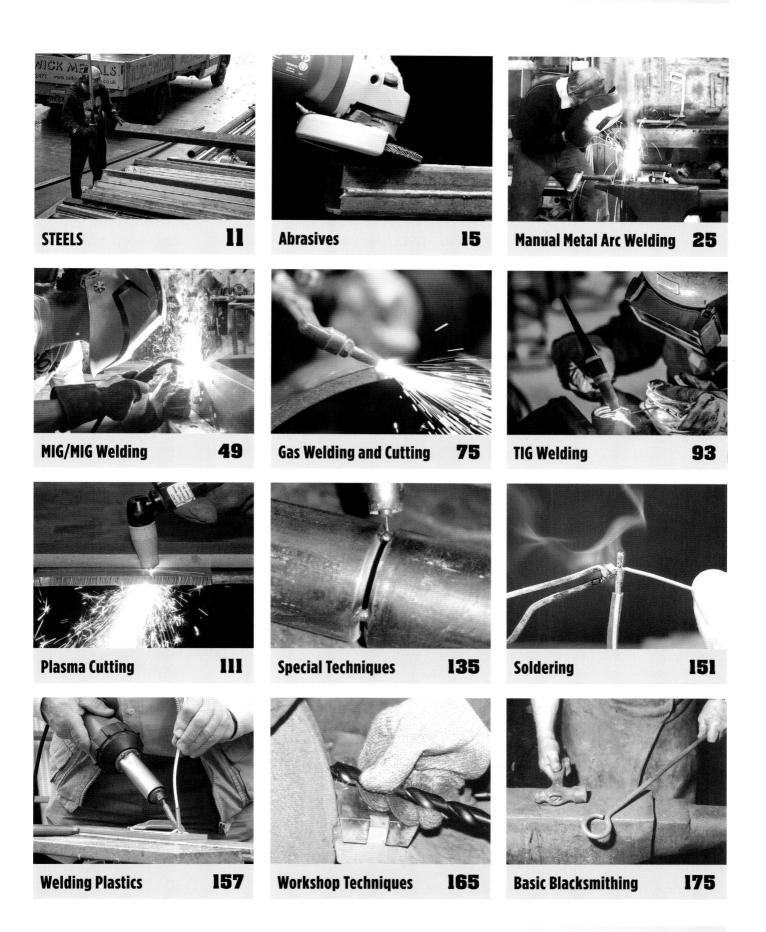

Introduction

Almost anyone can weld. All that's needed is basic hand/eye coordination, an idea of what's going on, and a little guided practice. So far nobody has come up with a book that grabs the reader's hands and says, "Do it like this!" And until someone does, pictures are worth a lot of words. So pictures are the heart of this book, forming a visual guide for beginners and a fault-finding service for improvers.

While not the last word on welding — other sources have a far better claim to that — the advice here is practical and aimed at the farm welder. The idea is to use the book for reference, dipping in and out to find snippets of info, the occasional hint, or maybe a way to get yourself out of a hole. It's not for sitting down and reading at one hit, unless you have a particularly high boredom threshold.

So what's coming? Twelve sections. First up is a run through types of steel and their uses. Abrasives, vital in pre-welding preparation and post-welding tidy-up, come next. The welding section opens with info on how to manually arc weld mild steel, detailing equipment, rod selection, plant setup, and work in various positions. As most people trip up on the same problems, there's a look at common faults and ways to fix them. Section 4 deals with MIG welding, a very inviting but unexpectedly complex process. Then it's the turn of the farmer's increasingly neglected mates: gas cutting and welding.

With the basics sorted, the trickier business of TIG welding steps up. Potentially the most versatile of all techniques, this can really expand workshop capability. Hard facing, pipe welding, and ways to join cast iron come next, before going back to the simpler jobs of how to solder and to weld plastics. The latter is not often done on farms but can be a big money-saver.

Rounding off are sections on common workshop skills — drill sharpening, the use of taps and dies and basic blacksmithing. How come blacksmithing? Knowing how to shape metal is a great help when making and fixing farm equipment.

It's pretty clear that reading can take you only so far. A good practical course is a fine way to improve, so check out your local college and training group. If that doesn't appeal, then YouTube has a mix of good and not-so-good videos. For my money the best online resource is Jody Collier's *weldingtipsandtricks.com* — check it out.

Should you wonder where the book's content comes from, it's based on the author's contributions to *Power Farming*, *Farmers Weekly*, and *Profi* magazines. My grateful thanks go to ESAB's Welding Process Superintendent Mick Andrews; to the Welding Institute's ex-chief instructor Max Rughoobeer; to plastics specialist Dave Tucker; to St Gobian (abrasives); to Rudgwick Metals (steels), and to Sussex blacksmith Frank Dean, who sadly died in 2004. Everyone's help was (and is) very much appreciated!

ANDREW PEARCE

Don't Take Risks

Here is the line taken on safety throughout the book. Today's duty of care requires that businesses and individuals both do, and are seen to do, the right thing regarding safe working practices.

So I don't intend to ramble on endlessly about staying safe. Adults should have enough common sense not to need that, and ought to pack a sufficiently well-developed sense of responsibility to look after the well-being of themselves, others, children, and livestock.

If a risk or hazard exists that might not be obvious, I'll try to point it out. Nevertheless, the responsibility to use good, sensible, and legal working practices rests entirely with the reader.

When welding, cutting, or grinding, use protective equipment — eye shields, fireproof clothing, proper footwear, and respiratory protection — as specified by industry guidelines. Advice on safe working is readily available: ask your welding equipment supplier, the equipment's maker, or contact Occupational Safety and Health (OSHA). Be aware of the presence of children and livestock, and the possible risk to them from hot material, fumes, flying sparks, and ultraviolet radiation. Assess whether your skills are up to the fabrication or repair that you're about to undertake, and think through the safety implications should it fail in service.

The bottom line is this: **if you're not 100% sure about how to do something and/or not 100% confident about the outcome, don't do it.**

First Things First

Just what is welding?

It's the process of joining materials using heat. In fusion welding, joint components are heated until they melt together or are positively fused by pressure. Blacksmiths use heat and hammer blows, but here we're more concerned with getting heat alone to do the work.

This heat will come from either an electric arc, a gas flame, or in the case of plastics, from a hot air gun. Filler is usually added to the joint from an electrode or separate rod. Non-fusion welding techniques like braze (or bronze) welding and soldering use heat too, but not enough to melt the metals that form the joint.

Metal Identification

This book deals mainly with welding mild steels. But as not all bits found under the bench or rescued from the scrapheap will be made of it, we'll start with different materials and their weldability. Although accurate identification of steel is a complex business, the main classes can be sorted out with a file, a grinder, and some basic ground rules. Section 1 expands on what follows, dealing specifically with steel grades and their application (see page 11).

Wrought iron isn't very common now, but has been used extensively in farming for chains and hooks. It's very low in carbon and malleable.

Mild steel is the common user-friendly stuff. It doesn't harden (much) when heated and cooled, and is easy to bend and weld. Black mild steel is what you'd normally buy: as flat strip and other sections it comes with radiused edges and retains its coating of mill scale from hot-rolling.

Bright mild steel in flat form has square edges, is shiny, and is more accurately sized than mild steel. It's made by cleaning and cold-rolling black mild steel, leaving the metal stronger but less ductile.

Silver steel looks like bright steel but is much harder. It contains chromium but, oddly, no silver. It's usually sold in short lengths.

Black and bright mild steels are easily filed and give off long, light yellow sparks under an angle grinder. Both are readily weldable. Silver steel is not.

Adding more carbon to steel makes it harder, and logically enough produces the **carbon steels** (Table 1). As carbon level climbs, so does the end product's hardness, brittleness, and difficulty of welding.

After forming to shape, carbon steels are often heat-treated (tempered) to boost their resilience. Welding heat can destroy the tempering effect, leaving the joint zone hard and brittle until it's re-treated. Springs are a classic case.

The more carbon in a steel, the harder it is to file — and files themselves have very high carbon content.

So here's a quick test. If an unknown material can't be filed, it's probably not weldable. The exception can be cast iron; see below. Grinding spark pattern also changes with carbon level. As it rises, the sparks get shorter, bush out closer to the grinding wheel, and may be darker yellow in color. If in doubt, compare sparks from the unknown item with those from a chunk of mild steel.

Although heat treatment will improve a carbon steel's resilience, the really spectacular gains come from adding small quantities of exotic elements to produce **alloy steels**. All sorts of metals — nickel, tungsten, manganese, molybdenum, cobalt, vanadium — can spice the mix, and the end result is usually heat-treated to maximize its properties.

Alloy steels turn up wherever toughness, resilience, and corrosion resistance is needed. Typical applications are springs, gears, and transmission shafts. Stainless steel is a variant using chromium to beat corrosion, which for the metalworker is both good and bad news. Although stainless steel is slow to tarnish, that

Two jobs using a dissimilar steels electrode: a sash cramp's cast iron endplate welded to the central mild steel beam for more rigidity (left), and a slurry pump's cast steel shear plate resurfaced back to near-original dimensions (right).

Table 1: Materials and their weldability

Material	Percentage carbon	Weldability	Typical use
Wrought iron	0.01–0.03	Good	Hooks, chains
Dead mild steel	0.1–0.125	Good	Wire, pressings
Mild steel	0.15–0.3	Good	General engineering
Medium carbon steel	0.3–0.5	May be made brittle	Structural steels, forgings, high tensile tube, some tools
High carbon steel	0.5–0.7	Will be made hard and brittle	Chisels, springs, hammers, railway lines
Very high carbon steel	0.9–2.0	Not weldable on farm	Files, razor blades, axes
Gray cast iron	2.0–4.0	Needs right method	Casting not subject to shock. Housings, pulleys, manifolds, etc.
White cast iron	2.0–4.0	Not weldable	Hard layer in wearing parts, camshafts
Malleable cast iron	2.0–4.0	Good with right method	Cutter bar fingers, coulter brackets, vice parts, clamps, etc.

reluctance to oxidize means it can't be gas-cut — but it can't resist a plasma cutter. And while many stainless steels are non-magnetic and weldable, don't weld if a magnet sticks to the bit you want to use; cracking is very likely.

Sorting an alloy from a carbon steel is largely a matter of application, though stainless stands out readily enough thanks to its satiny bright finish. Think about cost, too: a cheap hand tool is more likely to get its hardness from a tempered carbon steel than an expensive alloy one.

Castings can be recognized by their complex shapes, generally rough surface finish, and any raised surface lettering. But is the bit in your hand cast iron or cast steel? Application and a grinding test usually gives the answer.

Gray cast iron breaks very easily if bent or shocked to leave a grainy surface. But it stands compression loads very well, so turns up in machine beds, bearing housings, electric motor bodies, belt pulleys, engine blocks, manifolds, and such. Heat treating gray cast iron produces the much tougher **malleable cast iron**, which is close to mild steel in strength and ductility. Malleable cast is used where shock loads are high; in vice bodies, clamps, and pto shaft

yokes. **White cast iron** is very hard and brittle, properties that are used when a cast part must resist wear. So for some soil-engaging parts, the molten iron is chilled in specific areas while in the mold, forming an outer layer of hard while cast.

Cast steels stand much harder service, being tougher than cast irons and capable of being heat treated to boost their resilience. Cast steels turn up where a durable, complex shape is called for.

Telling the two apart is pretty simple. The quickest way is to grind them: cast irons give off unmistakable dull red/orange red sparks that don't sparkle and fade very close to the wheel, while cast steel sparkles clear yellow like

mild—though the sparks are closer to the wheel and more bushy.

The hammer test is another decider. Tap cast steel and it rings, while cast iron just makes a dull clunk. Other differences? Cast iron fractures to leave a very characteristic coarse, grainy, gray surface — break a bit to see — and if you drill or file it, the swarf is powdery. Cast steel produces silvery filings.

When you start to file or machine cast iron, it may seem very tough. This is down to a hard skin of white cast iron, formed on the surface where molten iron contacted cold sand in the mold. Break through this skin and the gray cast underneath files, drills, and machines very easily. Cast steels don't have this hard shell.

Iron and Steels: Carbon Contents and Uses

Blast furnaces make pig iron, which is high in carbon and impurities. Refining produces the following series of materials, with hardness and brittleness increasing as carbon content goes up.

Steels in the lower reaches of the carbon league are weldable on the farm. Ditto for those in the middle, though they need greater care in rod selection, joint preparation, and subsequent cooling. High carbon steels are unweldable by normal methods. Adding dashes of other elements to carbon steels gives a wide range of tougher alloy steels — see Section 2.

What Should You Weld?

Everything depends on the material involved and its application. Making 100% reliable joints in anything other than mild steel needs the right electrode and technique, and may call for specific procedures before and after welding to retain the metal's properties.

> There is only one rule. *Don't weld any safety-related component unless you're completely sure about its makeup and any heat treatment it may have had.* **If the part must be repaired rather than replaced, take it to a specialist.**

What are the options when safety is not at stake, or 100% reliability is not essential? Here a "dissimilar steels" electrode may be the answer.

Although metallurgists rightly stress the importance of matching rod and material, these jack-of-all-trades rods often get round material mismatches. If you're faced with joining carbon or alloy steel:

• Choose a rod that matches the most awkward of the metals to be joined.

• Preheat. A gas flame heats moderate-sized parts. Move it around to keep heat input even.

• Use the minimum current needed for fusion, and keep run number low.

After welding, let the work cool very slowly. Lay it on warmed firebricks or on dry sand, and cover it to keep off drafts. Don't put just-welded work on cold surfaces and never, ever, quench-cool. Even mild steel can harden a little if its carbon content is toward the upper limit, so where strength really matters, don't quench a mild steel repair in water.

Medium carbon steels can be stress-relived after welding by heating the joint area to very dull red and then cooling slowly (see distortion control, page 85).

Welding cast iron is covered in Section 4. Preheating gray castings helps a great deal, and low welding heat input followed by slow post-weld cooling are always necessary. Even then, success with cast iron is never completely certain thanks to the material's tendency to crack as it cools. It's important to know which cast iron you're dealing with: malleable cast will cool to brittleness if arc welded, so lower-temperature bronze welding is better. Gray cast will turn to the brittle white form if cooled too quickly.

Section 1

Steels

Mild steel is usually the first choice for making or fixing stuff in the farm workshop. And why not? It's easy to get, easy to work, relatively cheap, tolerant of heat/vibration, reluctant to harden, straightforward to weld, and easy to machine. In other words, a good and natural choice for most jobs. Yet it's not necessarily the best option. Other steels have properties that suit them better to specific applications — and by choosing material(s) with an eye to likely loads and service conditions, the final product of your efforts will most likely be stronger and more durable. An example: you need to change a pivot pin on an old bit of kit but the part is no longer listed. You could knock up a new one from mild steel bar, but depending on the load, what the pin runs in, and the lubrication arrangements, wear could be very rapid. But if you make the pin from a more suitable steel, it will last for years.

Qualified engineers know their steels and how/where to exploit them, with or without an accountant's eye to cost. For the rest of us it's handy to highlight some the many grades on offer and see

1.1. There's plenty of choice of steels and non-ferrous metals from a good stockholder.

their advantages and constraints. Before we kick off, a few points:

• Engineering steels are classified separately from structural steels.

• Engineering steels come in familiar sections — round, flat, bar, square, and hexagon — and include higher-carbon and alloy forms.

Structural steels are relatively low in carbon and come as beams and columns (RSJs), as well as in channel, flat, round, square, sheet (including galvanized), mesh, and finally tube (round and box, seamed or seamless).

• Engineering steels come in various finishes. The most common are black

1.2. Bright steel is cold-drawn to size, which produces a shiny finish. This is EN19T.

1.3. Continuously welded box or tube has a seam. This projects inwards, making life difficult when you're trying to get one section to slide freely inside another.

1.4. Seamless box or tube has no internal weld projection and is more uniform in size, so is stronger than the equivalent welded section. Cold-drawn seamless (CDS) tube comes in different grades; some can readily be bent, others are more rigid.

1.5. Black steel, whether engineering or structural, is hot-formed and carries some mill scale.

1.6. Steel comes in standard lengths and is usually color-coded. But as there is no standard code, individual mills and retailers can (and do) use their own scheme.

1.7. Key steel is a medium-carbon, bright material finished to a wide range of sizes. Depending on the application, substituting a softer mild steel version of a damaged key may cause early failure.

(hot-formed with residual mill scale) and bright (cold-formed, shiny finish). Ground bar steel is bright and finished to close tolerances, so suits shafts. Structural steels are black-finished.

• The supplier will be your friend if you list the planned use, type, section, size, length, and finish of material before calling. This saves time and is a basis for discussion.

• Various standards and grading systems are used to classify engineering steels. In this chapter the older BS970 and the newer BS 1991 reference

numbers are used, as these are widely recognized in the UK. Structural steel BS EN grades are prefaced by "S."

• Steel may be color-coded for identification, but the system depends on the mill and/or supplier. There is no UK standard.

• Carbon content, alloying element percentage, and tensile strength figures vary slightly between suppliers.

• The following examples are either carbon or alloy steels. Stainless steels, tool steels, and non-ferrous metals have their own grades.

• On page 13, tensile strength is given as both maximum and yield point strength values.

A caution before going on. While mild steel is very tolerant of the usual forming and joining processes, other steels may not be. Heat, either used to bend the material or from welding, can substantially alter their properties. Cold bending may make the material brittle; heating can soften or harden it, depending on cooling; welding may need specific consumables and/or techniques. Before buying, sort out exactly what you want to do with (and to) the steel. Then discuss which material will be most suited to the application with a good supplier. *Take professional advice if you're not 100% sure.* That way the result will be safer and more cost-effective.

Carbon Steels

Steel is an alloy (a mix) of iron and other elements. Carbon steels are ranked according to their carbon content, though this is not the only extra element in the mix. Carbon content typically runs from 0.15% max to 2.0% — definitions vary — with 0.5% usually seen as the upper limit for medium carbon material, though that definition too can be hazy. The more carbon in the steel, the harder and tougher it can be made by heat treatment but the trickier it is to weld. Mild steels (and there's more than one) contains up to 0.25% carbon. Here are some common carbon steels in ascending content order.

BS 970 EN1A

Equivalent BS 1991 grade: 230M07

Carbon content: 0.15% maximum

Max/yield tensile strength: 360/251 N/mm^2

Sold as black or bright in round, square, flat, hexagon.

A low carbon, free-cutting mild engineering steel for machining in manual or automatic lathes. Swarf forms small chips. Not readily hardened.

BS 970 EN3A

Equivalent BS 1991: 270M20

Carbon content: 0.16%–0.24%

Max/yield tensile strength: 400-560/300-440 N/mm2

Sold as black or bright in round, square, flat, hexagon, angle; also ground steel bar

The most common of the low-carbon mild engineering steels, used for general fabrication. Easy to bend and readily welded with MIG, MMA, TIG, or gas. Not good in high-strength applications. Can

be case hardened if heated and quench-cooled. May then be tempered, but EN8 or EN9 are preferable for this.

BS 970 EN8

Equivalent BS 1991: 080M40

Carbon content: 0.35%–0.45%

Max/yield tensile strength: 700-850/465 N/mm2

Sold as black or bright as round, square, flat and ground bar. Price premium roughly 10% over mild steels.

A medium carbon, medium tensile strength machinable engineering steel. Readily hardened and tempered to improve wear resistance. Use where mild steel won't do but where an expensive alloy steel is not justified: axles, some pins, studs, shafts. Watch out for hard, brittle face potentially left by flame cutting. Weld with MIG or low hydrogen MMA rods. Preheat thick sections to minimize change of cracking.

BS 970 EN9

Equivalent BS 1991: 070M55

Carbon content: 0.50%–0.60%

Max/yield tensile strength: 600-700/310-355 N/mm2

Sold as round, square, flat, plate.

More carbon than EN8 but still a medium carbon engineering steel. Readily heat treated, resists wear well. Often used for sprockets, gears, and cams. Pre/post heat and specific consumables needed when welding.

BS 970 EN43 Spring steel

Equivalent BS 199: 080A57

Carbon content: 0.45%-0.60%

Max/yield tensile strength: 380/210 N/mm2

Sold as bar and plate

A carbon steel with manganese and silicon for oil hardening and tempering. Used for springs and hand tools.

BS EN 10025: S275

Carbon content: 0.25% max

Min yield tensile strength: 275 N/mm2

Sold as beams, columns, flats, tubes, etc.

Use: A low-carbon structural manganese steel for general fabrication and building work. Easily welded by common processes. Name reflects its minimum yield strength — 275 N/mm2.

BS EN 10025: S355

Carbon content: 0.20% max

Min tensile strength: 355 N/mm2

Sold as beams, columns, flats, tubes, etc.

A low-carbon structural manganese steel with slightly different composition than S275. Easily welded by common processes. Better impact resistance than S275, easily machined, better in demanding environments. Name again reflects minimum yield strength — 355 N/mm2.

BS46 Key steel

Carbon content: 0.40%–0.45%

Max tensile strength: 500-700 N/mm2

Sold as bright squares and flats

Medium carbon steel drawn to specific tolerances in metric and imperial sizes. Harder than mild steel, used for square, taper, plain, half moon and gib head keys.

Alloy Steels

Alloy steels result from adding various proportions of extra elements — typically manganese, chromium, nickel, boron, vanadium, and molybdenum — to medium carbon steels. The result is a wide range of engineering metals that are intrinsically tougher and more resilient than carbon steels, and whose properties (including resistance to impact, wear, and corrosion) can be extended by heat treatment. For example, case hardening or nitriding produces a material with a hard exterior and resilient core. Alloy steels are relatively expensive, vary in machinability, need consideration and care when welding and come in only a few sections, limiting their use in the farm workshop. Components made from them are often machined, heat treated, and finally ground to exact size. Unless you really must have properties that only an alloy steel can bring, then a medium carbon steel like EN8 — perhaps heat treated by a specialist to match the application — is often enough.

BS 970 EN16T

Equivalent BS 1991: 605M36

Carbon 0.36%, Mn 0.45%, Mo 0.20%, Cr 1.00%, Ni 1.30%, Si 0.10% (min values)

Max/yield tensile strength: 850-1,000/680 N/mm2

Supplied usually heat treated as round, ground bar, hexagon, black or bright.

A ductile, shock resistant, low alloy manganese-molybdenum engineering steel. Machinable. Usually used for high strength, resilient shafts and axles; also bolts, cams etc.

BS 970 EN24T

Equivalent BS 1991: 817M40T

Carbon 0.36%, Mn 0.45%, Mo 0.20% , Cr 1.00%, Ni 1.30%, Si 0.10% (min values)

Max/yield tensile strength: 850-1,000/650 N/mm2

Sold as round, square or flat, black or bright.

Widely used nickel-chrome-moly engineering steel combining strength, wear resistance, shock resistance and ductility. Can be heated to extend its properties in several directions. Used for driveshafts, shafts, gears, cams, etc.

BS 970 EN45 spring steel

Equivalent BS 1991: 250A53

Carbon 0.5%, Mn 0.7%, Si 1.50%, p 0.05% max, S 0.05% max

Max/yield tensile strength: 1,500/1,100 N/mm2

Supplied as round and flat.

Common in vehicle applications. Very resilient spring characteristics once oil hardened and tempered. Used for the making and repair of leaf, coil and flat springs.

The above examples cover the popular ground. When you aren't sure which steel is right for a given job, the best thing is to speak to a competent supplier. With the material sorted out, the stockholder should be able to point you (if necessary) to advice on welding process(es) and/or heat treatment. Take that line with your steel choice and you'll end up with a better result — not to mention a potentially cheaper and safer one — than if you just use whatever you can find and hope it will be OK.

Section 2

Abrasives

Whenever you set out to weld, braze, or solder, cleaning the work is number one priority. And with steel, the easiest, fastest, and most effective route to get there is via one or more abrasives. There's a wide choice out there, in form (cloths, discs, flap wheels, stones) and in method of using them, from manual work to power tools. Once the latter swing into play they bring a risk element, too — so for starters, here's something to remember: *angle grinders and their bench-mounted brethren can do you serious harm if not treated right.*

Given that encouragement, we'll sort out safety first. Mostly we use grinders on steel and on stone without a second thought, but ponder this. Spinning without load, the outer edge of a typical angle grinder disc flies by at 80m/s. That doesn't sound like much. Yet 80m/s is 290km/hr, or a scorching 180mph. And if you're in any doubt over what can happen when a disc bursts, just search the web for "angle grinder accident." So here's how to avoid that.

A good-quality abrasive wheel or disc is made and tested to very strict standards and is extremely unlikely to give the user any grief — but only when the wheel is chosen, mounted, and used properly. Humans being humans, accidents tend to start in the last part of that.

We're dealing here with bonded wheels. That is, types where the abrasive is held in a solid matrix or binder. The nature of the binder and the type of construction then splits bonded wheels into two broad classes — **resinoid** grinding/cutting wheels (or discs) used mainly in handheld tools, and **vitrified** wheels used in bench grinders. The two classes have very different properties and need very different treatment. There are also **flap wheels**, made from overlapping abrasive strips — more on those later.

A word on quality and its impact on safety and service life. In Europe, all bonded wheels must conform to EN12413. Among many other things, this requires each wheel to show the relevant standard and to carry vital information on its maximum permitted peripheral (outer edge) speed and rpm. Good-quality wheels always carry this info, along with pictograms laying out the wheel's application(s), recommended safety equipment, and markings that allow the wheels' origin to be traced. Look for these, plus the oSa (Organisation for Safety of Abrasives) symbol — see "Wheel markings" for more. Lower-quality wheels are usually cheaper and may not have all of the above. They can also be much more variable in quality and will probably have a shorter service life. Abrasive flap wheels must conform to a different standard — EN13743.

For material removal and cutting metal or stone, thin **resinoid-bonded wheels** — usually angle grinder discs on the farm — are the thing. Ranging in thickness from 1/32" to 3/8" (1mm to 10mm), they hold abrasive grit in a resin matrix, and always feature fiberglass mat reinforcement. The resin and mat bring toughness, allow high-rpm operation and in thinner discs, allow significant flexibility. Resinoid wheels are built up from layers of bond and mesh, pressed in a mold and cured at relatively low temperatures; a metal center disc usually provides an accurate mounting surface. For typical farm-sized 4", 4½", 5", 7", and 9" (100mm, 115mm, 125mm, 180mm, and 230mm) angle grinders, the maximum peripheral wheel speed is 80m/s, regardless of wheel diameter.

Vitrified bonded wheels are quite different. Designed for use on bench grinders, they use a mineral or synthetic material to hold the abrasive. Fired at very high temperature, this turns into a hard, brittle, and glass-like bond. As there is no separate reinforcement, vitrified wheels are more prone to shock damage than resinoid wheels and can't stand high rpm. So their maximum peripheral speed is lower at 35m/sec-50m/s.

Wheel care and machine use comes down to six golden rules. Stick by these, add a healthy dose of respect for the equipment and you'll be fine.

The rules:

• Match the wheel to the work.

• Be sure that the machine can't physically spin the wheel faster than its maximum allowable peripheral speed or rpm (see "Stumped for speed").

• Mount the wheel properly.

• See that machine guards are in place and secure.

• Keep the wheel's design in mind while using the tool — for example, don't grind with a cutting disc.

• Wear the right protection for eyes, hands, ears, and lungs.

A full safety code can be found at *www.fepa-abrasives.org.*

Types Explored 1: Resinoid Cutting and Grinding Wheels

For safety and good service life, it's essential that discs for grinding and cutting are intact, spin at the right speed, and be mounted securely. Beyond that, how you present the wheel to the work has a big impact on material removal speed as well as safety.

And of course the machine's guard must be fitted. Some of the nastiest accidents happen after a guard has been taken off, usually for better access to a corner or to fit the wrong size of wheel. Such daftness aside, changing wheels and using a handheld grinder comes down to common sense, though not everything is obvious, as the pictures below show.

2.1a and b. See that the machine you use can't spin the disc faster than its maximum allowed speed. Find the disc's max rpm (2.1a) or calculate it if it's not visible — the box "Stumped for speed" shows how. Here it's 6,650rpm. Then find the grinder's max spindle rpm (2.1b); in this case 6,600rpm. As that's 50rpm lower the wheel's limit, the pairing is safe to use. Last, check out the disc for cracks and/or missing bits before you mount it, and junk it if you find either.

2.2a and b. Not everyone notices that the upper spindle flange on angle grinders is generally reversible. The side with a raised inner rim (2.2a) is for use with grinding discs. Fit it so the raised section locates and centers the disc. Use the flat and/or recessed side (2.2b) with cutting discs. Get the flange the wrong way round with a cutting disc and the slim raised rim limits clamping force; the disc can then slip and may stall in the cut, with a risk of damage to you and it if you just yank it out with the tool under power.

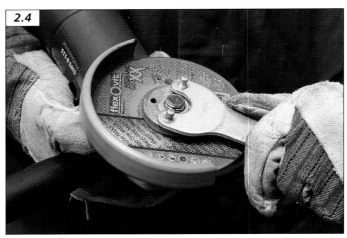

2.3. Discs with a depressed center must be fitted with the low section toward the machine body, as the depression adds mechanical strength. Don't fit a disc the wrong way up even if the tool allows — it's not designed to run that way.

2.4. Switch off the grinder at the mains and unplug it before changing wheels. Go easy when tightening the flange; just nip it up enough to grip the wheel, no more, then spindle rotation will build on that. Over-tightening means struggle when you come to change discs.

Wheel Markings

A good-quality disc carries all the info you need to suit it to the job, match it to the machine, and use it safely. Pictured is a 9" (230mm) angle grinder cutting disc that, like all good-quality products, has info permanently etched in during manufacture rather than added as a sticky label. First thing to look for is the EN12143 mark, which on this wheel is shared with a unique identifying barcode (A). Next find the wheel's max permissible speed, here given in m/s and rpm (B). Also here are the wheel's dimensions in mm and inches, plus a code — A46T-BF41. Use Table 2.1 to crack these codes: this one breaks down as follows:

- A = abrasive type (aluminum oxide)
- 46 = grit size (46 per linear inch)
- T = grit hardness (hard)
- BF = bond type (Bakelite/glass fiber reinforced)
- 41 = shape specification (flat cutting-off wheel)

Item C in the picture is the range of applications the disc is designed for. Also obvious should be a selection of symbols (D). These point to the safety equipment the user should wear and show specific details, for example that the wheel contains no iron, sulfur,

Wheel markings

or chlorine (so won't contaminate stainless steel) and that it must not be used for grinding. Item E is the oSa symbol, confirming that the makers belongs to the Organisation for Safety of Abrasives; this guarantees that the product meets at least EN 12413.

Finally, look at the center ring. This should show the wheel's validity date.

The organic bonding in resinoid wheels degrades over time, so a three-year shelf life is usually recommended. The ring should be stamped with the use-by year, along with a two-digit code showing which quarter it was made in. In this case, the expiration date was the third quarter of 2012.

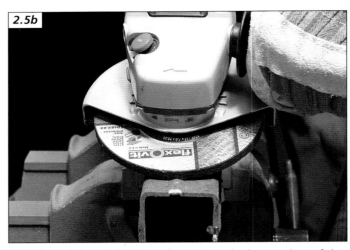

2.5a and b. An angle grinder's guard can — and should — be rotated to suit the job. In 2.5a it's set for cutting: the lower edges of the guard line up with the grinder's body. Picture 2.5b shows the guard turned through 90° for grinding. The important thing is that the guard always sits fully between the wheel and you.

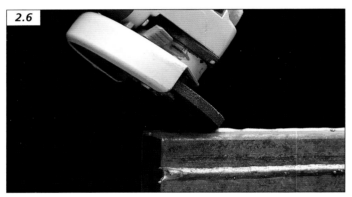

2.6. Run grinding discs at 30°–45° to the work. A relatively small contact patch then removes material fast. Shallower angles spread the wheel's pressure and slow down material removal. Low angles also thin the edge of the wheel, which often then loses small chunks and becomes ragged — usually as wheel diameter decreases.

2.7. Flap discs need to run at a shallower 10°–15° angle to put maximum abrasive in contact with the work.

2.8. Cutting wheels are thin, partly to concentrate pressure on the working edge and partly to minimize cut width. Although reinforced, these wheels are definitely not designed to take side loads — it's very easy to snap a 4½" x ⅓₂" (115mm x 1mm) item with just finger pressure. Bring this picture to mind every time you're tempted to grind with a cutting disc.

2.9. It follows from 7 that a cutting disc should be used vertical to the work, as shown. Load then passes safely up through the disc. Side loads can fracture a disc wheel or tear out the center, so trying to cut in anything other than straight lines is likely to bring grief.

Table 2.1: Codes Decoded
Here's a roundup of standard markings found on resinoid angle grinder discs.

Abrasive type	Grit size	Grade	Bond				
A = aluminum oxide C = silicon carbide Z = zirconia alumina BF = Bakelite/glass fiber	Coarse	Medium	Fine	Soft	Medium	Hard	B = Bakelite/resinoid*
	16	30	80	N	Q	T	
	20	to	to	O	R	U	
	24	60	100	P	S	V	
Shape							
BF 27 = depressed center, grinding							
BF 41 = Flat cutting							
BF 42 = Depressed center, cutting							
Validity date (on center ring)							
V01 = made in first quarter							
V04 = made in second quarter							
V07 = made in third quarter							
V10 = made in fourth quarter							
Note: The quarter code is followed by the wheel's recommended use-by year. *Not for use on angle grinders — not reinforced.							

Flap Discs

Resinoid discs are the roughnecks of the angle-grinding world. We've seen that the grinding variety is best used at a relatively steep 30°–45° angle to the work, which concentrates your downward pressure to make the abrasive cut. Resinoid discs can handle any material apart from aluminum and other soft metals that clog the abrasive, but they do leave an uneven finish. On heavy jobs like removing old weld metal, scouring out rust pits, or cleaning up gas-cut edges, that's not a problem.

Resinoid discs cover most bases. Good-quality ones aren't expensive and offer sensible service life. On top of that most farms work predominantly with steel, and getting a newly welded fabrication or repair out of the door usually takes priority over the job's appearance — but sometimes you need something a little more genteel. That's where flap wheels step in.

Pictures show items from Norton's range, but other brands have comparable products. Expect to pay a premium for high-end items, but expect the reward of better performance and longer life.

Where can you use flap discs? At the three stages stage of repair and fabrication — surface prep, metal removal, and cosmetic finishing.

1. Surface preparation. Here the broad class of **non-woven discs** is good at getting shot of paint, mill scale and light rust, along with platings like zinc galvanizing and other corrosion-preventers. Rather than holding an abrasive in a rigid and reinforced resin, non-woven discs put it in a web of synthetic fibers and hold it with a dry smear-resistant adhesive. Various types and grades of abrasive can be held in the web, which is supported by a thin, consumable backing plate. By adjusting the type of abrasive and its grit size, makers generate a family of non-woven discs able to grind at one extreme and polish

Stumped for Speed?

How do you find the safe maximum rpm that a wheel can be spun at, if all you can see is its max peripheral speed? A little recreational math is called for:

$$\text{Max peripheral speed (m/s)} = \frac{\pi \times \text{wheel diameter (mm)} \times \text{rpm}}{60{,}000}$$

So for a wheel of 230mm (9in) diameter and a max peripheral speed of 80 m/sec:

$$80 = \frac{3.142 \times 230 \times \text{rpm}}{60{,}000}$$

Which rearranges to

$$\text{Rpm} = \frac{80 \times 60{,}000}{(3.142 \times 230)}$$

So in this example, safe maximum rpm = 6,642.

at the other. Non-woven discs do a fine job removing light rust, paint, scale, and other weld contaminants from metal, plus they can strip unwanted crud or paint from cement or even wood. Some types can handle relatively heavy grinding work as well, depending on the pressure applied, and are can even work on rough, snaggy material. But naturally around such surfaces, the softer the disc, the more care that you need to take to avoid damaging it.

A non-woven disc is typically ½" (12mm) deep, that is, front face-to-back face. When the front face is used at a shallow 15°–20° angle, its generous area lets you cover a lot of ground quickly. Or where access allows, you can use the outer rim in the angle of a joint for a final clean before welding. Negatives compared to resinoid discs are more-or-less sensitivity to sharp edges, a shorter service life, and higher cost/life ratio.

2. Metal removal. Usually this involves tidying-up joint edges before welding and if necessary, beveling them. The ability to grind a decent bevel also comes in handy when sharpening hedge cutter flails and mower blades. Resinoid discs will do the job, but producing a flat, uniform bevel is a bit of a lottery. Instead, a **flap disc** produces a much better finish and is more controllable. Flap discs are made from overlapped strips of abrasive cloth anchored to (and supported by) an angled fiberglass backing plate. The plate's angle is a shallow 15°, which automatically suggests the disc should be used at a similar low angle; this in turn produces a big footprint on the work and lets the flap disc deliver a flat, uniform finish. Beyond grinding bevels, flap discs find plenty of use in blending or tidying metal, de-burring edges and holes, or on any other work where their forgiving, contour-following nature can be exploited. The downsides? You can't grind with a flap disc's edge, service life is shorter than for resinoid types, the per-item cost is higher, and the flat footprint makes cleaning out rust pits more difficult. Horses for courses, as they say.

3. Finishing. Where time allows it's always satisfying to bless a job with a good cosmetic finish. Not a mirror polish (though that's possible) but knocking off weld spatter and blending-in bead edges, then perhaps stripping off all old coatings on repaired parts before repainting. Here a flap wheel and non-woven discs can work as a team: a flap wheel for initial metal removal and a non-woven one(s) for final blending and stripping. This suggests each has its own exclusive territory, though in reality their applications overlap. In common their big contact area means fast work, and as ever with abrasives, smaller grits deliver a finer finish — see "Graded Grains."

Graded Grains

By juggling the abrasive and bond, resinoid and vitrified wheels are produced. Within these the bond strength, the abrasive type and the grit size suits the disc or wheel to different work.

Resinoid cutting and grinding wheels use several different abrasives. Two are very common and suited to farming applications:

• **Aluminum oxide (alumina).** A general-purpose abrasive for grinding and cutting metals only. Will not cut stone. Naturally occurring or synthesized.

• **Silicon carbide.** Recommend for stone but will also cut most steels. Sharper grains than aluminum oxide.

With either, grit size choice is narrow and not too important in general farm work.

Good-quality resinoid wheels carry a hardness grade letter — see Table 2.1. Grades show the relative holding power of the bond: soft grades have earlier letters in the alphabet and release abrasive earlier, so exposing fresh sharp grains.

Vitrified wheels give a wider choice of abrasive. The most common are:

• **Aluminum oxide.** Color white or brown; general-purpose use.

• **Silicone carbide.** Color green. For harder materials including high-speed steels and tungsten carbide tooling, non-ferrous metals and cast iron.

• **Ceramic.** Color blue. For hard-to-grind materials. A long-lasting high-performance material.

• **Alumina-zirconia.** Comes from the fusion of aluminum dioxide and zirconium dioxide, toughening the alumina. Extremely hard grains with high resistance to dulling.

Grit size choice is also wider in vitrified wheels, running from 24 (coarse) to 120 (fine). Which to buy depends on the application: coarse grits give faster material removal, finer grits give a finer finish. One option is to mount coarse- and fine-grit wheels on one bench grinder, say a 46 grit on one side and a 60 grit on the other. Alternatively, fit an aluminum oxide wheel and a silicon carbine one so the one grinder can handle a range of materials.

In general, softer-bond wheels are best for hard materials like drills, lathe tools, chisels, and so on, along

The Safe Angle

Flap- and non-woven discs are slightly less threatening than the resinoid variety. Although spun at the same rpm by the grinder, fragments from a non-woven disc generally have less mass and so lower impact energy. That's not to say that the basic rules of angle grinder use can be ignored, though, so . . .

• Mount the disc securely using the right flanges.

• Use the grinder's guard, set so it's completely between the disc and you.

• Wear eye protection and a mask that protects against abrasive/matrix /material dust.

Practicalities

For best performance, 4½" (115mm) flap or non-woven discs with ceramic-based abrasive should be used with a 1,000W–1,500W angle grinder. Why? Lower-powered machines won't necessarily put enough energy into the abrasive to fracture its grains (a process necessary to create fresh cutting edges) so the disc can't perform as intended. Other considerations:

• As non-woven discs are substantially thicker than a conventional dished item, you may need to add a cranked section to the grinder's pin spanner so it can reach the spindle flange.

• Disc depth means that a quick-release guard can make it easier to change non-woven discs

• Non-woven and flap discs have less dense, more flexible base material than resinoid ones. So they follow contours better, feel less harsh to use and generate less vibration — but they do shed many tiny flecks of matrix as they wear, so post-work cleanup needs to be thorough.

How Does an Abrasive Wheel Cut?

The way discs and wheels cut or grind is pretty nifty. Abrasive grains — which today are often synthetic — are held in the wheel's bonding matrix. Each grain is like a small chisel, able to slice off slivers of material. Like any cutting tool, a grain loses its edge after a while so starts to rub more than it cuts, so friction goes up. The grain's temperature climbs until it melts the holding matrix, at which point the spent grit is flung free and fresh, sharp abrasive is exposed. Clever, eh?

with large contact areas and rapid material removal. Harder grades are more general-purpose and are better for softer materials, on small contact areas and where longer service life is important. Medium-hardness wheels suit most farming applications.

Vitrified Wheels

To round off this section we'll gallop through the third major type of abrasive wheel — those used with bench grinders and low-low-powered craft tools.

As we said earlier, vitrified wheels hold their abrasive in a mineral or synthetic material. During manufacture the wheel is fired in a kiln at very high temperature, which turns the material into a hard, brittle and glass-like bond. This leaves vitrified wheels more prone to shock damage than resinoid discs discussed last time, and more

rpm-sensitive. Their relative fragility means you must take particular care in mounting them and in use, and key points are covered in pictures.

It's essential to make sure that the grinder can't spin the wheel(s) too fast. Both the wheel and grinder should have info on their respective speeds (peripheral/rpm on the wheel, spindle max rpm on the grinder), and as before the grinder must spin more slowly than the wheel's allowable maximum. *If you can't be sure that the wheel will be working at a safe speed, don't mount it!*

And it's doubly important to check a vitrified wheel's physical integrity before you fit it. Look for cracks or damage, and do a ring test. The highest risk of bursting or fragmentation comes each time a wheel accelerates from rest, so once a new wheel is mounted,

Can I . . .

. . . grind on the side of a vitrified wheel? Tempting, especially if the face of the wheel is not flat and you need to sharpen a drill or a lathe tool on it. But it's risky: conventional flat vitrified wheels are not designed to stand side loads and can shatter. Instead, dress the wheel's proper working face flat and use that (see photos 2.16a, 2.16b, 2.17).

. . . use a worn-down 9" (230mm) disc in a smaller angle grinder? Another temptation to avoid! Spindle speeds on smaller grinders must be relatively high to produce the right peripheral speed on a small-diameter wheel. So there is a real risk of over-speeding a worn larger 9" (230mm) wheel, even though its diameter is much smaller than when it was new.

stand clear on start-up and run it for a minute or two to make sure all is OK. You should treat a vitrified wheel with particular care when grinding; make contact with it gently and let the wheel do the work. Don't force material in, and keep the work down on the grinder's tool rest.

Again stick by the six rules:

1. Match the wheel to the job.

2. Be sure that the machine can't spin the wheel faster than its maximum allowable peripheral speed or rpm.

3. Mount the wheel properly.

4. See that machine guards are in place and secure.

5. Respect the wheel's design capability in use.

6. Wear personal protection for eyes, hands, ears, and lungs.

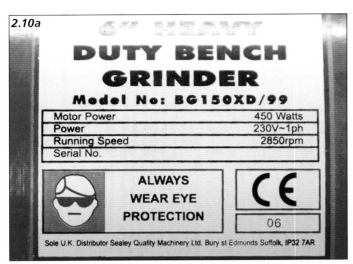

2.10a and b. A bench machine's info plate should show spindle rpm (2.10a); measure it using a tacho if not. Likewise the wheel should carry its maximum peripheral speed and/or rpm (2.10b). If the wheel won't be spun faster than its design speed and is physically in good shape (see 2.11), then go ahead. If you can't find the wheel's maximum speed, DON'T USE IT!

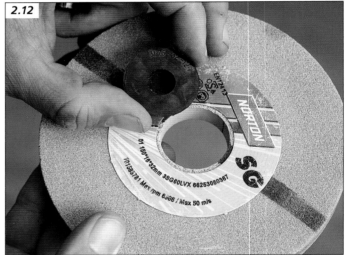

2.11. Vitrified wheels can stress-crack if not stored and handled with care. So keep them dry, laid flat or stood on edge. Surprisingly they are porous and will take up water, which can unbalance the wheel and increase the risk of bursting. So before fitting any vitrified wheel, check it visually for water stains, cracks and chipping. Then support it on a finger and tap it with a length of hardwood, not a metal tool. A good wheel will ring, a damaged one will produce a dull clunk. The ringing won't be like a bell but you'll recognize it.

2.12. Bench grinder wheels commonly have a center bore of 1¼" (32mm or 31.75mm). This is much bigger than the spindle of smaller machines, so bushes are needed to match one to the other. Usually plastic, these are slightly tapered, come in pairs and are readily available. It's essential that the bushes are a snug fit in the wheel. Don't use Imperial ones in metric wheels as they are slightly too small; the resulting slack lets the wheel run eccentrically, greatly increasing the risk of bursting. Metric bushes are OK in Imperial wheels as their taper accommodates the size difference. Also make sure that the wheel's two soft paper blotters are in place. These cushion the wheel against surface roughness, which will otherwise act as stress raisers (see photo 2.13).

2.13. Vitrified wheels are sensitive to point pressure. This can be the source of a dangerous stress fracture in use. Before mounting a wheel, check the spindle and both flanges for grit or metal fragments that, if trapped against the wheel, will concentrate stress (arrows).

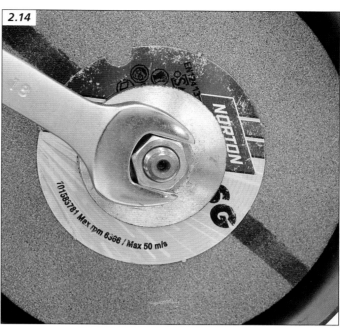

2.14. Unplug the grinder before working on it. Tighten the spindle nut just enough to secure the wheel snugly: it will self-tighten in use.

2.15a and b. Plenty of accidents come from tool rests that are not set close enough to the wheel. The gap in pic 2.15a is too big — material can be pulled into it, risking damage to the work, wheel, and operator. Set a minimum clearance gap to a new wheel, then as the wheel wears down keep adjusting the tool rest to maintain the gap at less than ⅛" (3mm) (2.15b). If you can, set the tool rest on (or a little below) the spindle centerline to minimize the tendency of the work to pull into the clearance gap. When using a wire brush in place of a wheel, it can be safer to remove the tool rest.

2.16. Should a wheel burst, it will cause minimum havoc if the guards enclose as much of the flying material as possible (A). Use the grinder's own eye shield (B). If that's too scratched, switch to safety glasses or goggles. Grinding without eye protection is only for fools.

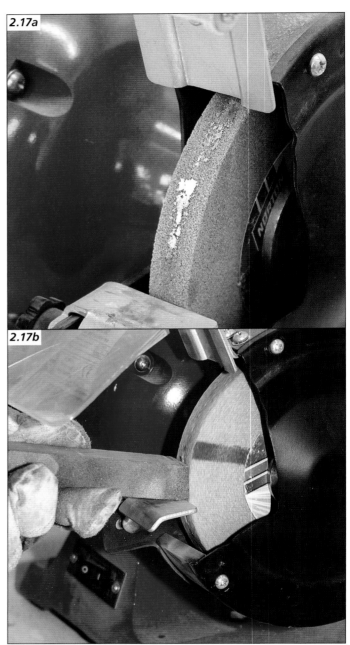

2.17a and b. Vitrified wheels can clog with soft material or can glaze over (2.17a). Then the only way to expose fresh grains is to use a dressing tool: the cheapest and easiest option is a dressing stick. Available from engineering suppliers, the stick must be sharper-grained than the wheel if it's to work. So for example, choose silicon carbide for an aluminum oxide wheel. Use the stick as shown in 2.17b. Dressing is also an informal way to bring the face of a wheel back to flatness before sharpening a drill. The softer bond of resinoid angle grinder discs means they self-clean, so don't need dressing.

2.18. A star wheel dresser exposes fresh grains, takes out minor eccentricity from the wheel to cure vibration, and/or reshapes the grinding surface for specific jobs. Available in different sizes, they use replaceable hardened, serrated, free-running wheels to dislodge old abrasive. Dressing is a dusty job, so be sure to wear protective gear and cover anything close by that matters. Be sure to re-adjust the grinder's tool rest when you're done!

Manual Metal Arc Welding

First Things First

Just what is welding? It's the process of joining materials using heat. In **fusion** welding, joint components are heated until they melt together or are positively fused by pressure. Blacksmiths use heat and hammer blows, but here we're more concerned with getting heat alone to do the work.

Heat comes either from an electric arc, a gas flame, or in the case of plastics, from a hot air gun. Filler is usually added to the joint from an electrode or separate rod. Non-fusion welding techniques like braze (or bronze) welding and soldering use heat too, but not enough to melt the metals forming the joint.

Manual Metal Arc Welding

Manual Metal Arc (MMA, or "stick") welding is still probably the most common technique on the farm. Best suited to materials thicker than ¹⁄₁₆" (2mm), it scores through equipment portability, a relative insensitivity to wind when working outside, and a relative tolerance of contamination. These relativities will become clear

later on. Stick welding is not the fastest technique but done properly, produces strong joints in a range of thick materials.

What makes a good fusion weld? Pictures 3.1-3.3 show the basics. Achieving **fusion** and **penetration** is the key to sound work in any process: fall down on either and joint strength

will be lacking. Four runs (or passes) went into picture 3.2. See how weld metal smoothly fuses into the original (or parent) plates, how it penetrates through them, and how the individual runs blend together? The result is plenty of strength in the joint.

Picture 3.3 shows what's happening during a good MMA weld. An arc

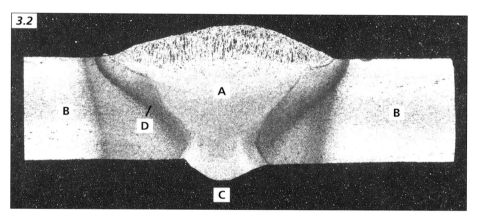

3.2. Sectioning the joint in (1) shows how weld metal (A) has fused with the two parent plates (B) and penetrated through them (C). Good fusion and penetration underpin every strong, durable weld. Notice how metal is built up slightly at the top and bottom of the joint to give a convex profile. And a smooth blend at the surface where weld and parent metals join avoids stress concentration, reducing the chance of failure.

3.1. A solid weld — what you're aiming for. Spot the even U-shaped ripples, full fusion at the edges and the lack of spatter? Good welds come from getting right the four variables (page 34). Note how only the top pass retains a coarse grain structure. The ones below it have been refined and strengthened (normalized) by the heat of each subsequent run. Thus several smaller welds are usually stronger than one big one.

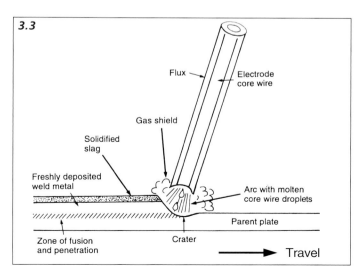

3.3. This is what was going on at the rod tip as weld 3.1 was made.

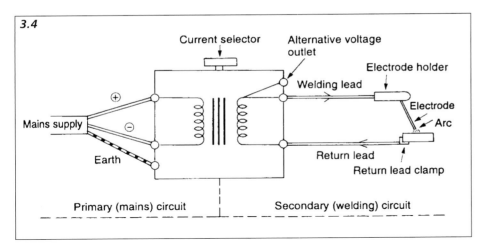

3.4

Current selector — Alternative voltage outlet — Electrode holder — Welding lead — Electrode — Arc — Mains supply (+) (−) — Earth — Return lead — Return lead clamp — Primary (mains) circuit — Secondary (welding) circuit

3.4. The conventional MMA set is based around a transformer. High voltages and low currents go in (left); low voltage(s) and high currents come out (right). Welding current flows out of the transformer to the electrode holder, down the electrode, across the arc to produce welding heat, and back via the return lead (arrows). While it's tempting to call the return lead an "earth," that term applies only to the incoming supply's ground wire. Keeping this in good order should keep the user safe in case of short circuit in the mains supply (see 3.5).

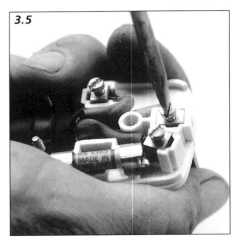

3.5

3.5. Mains supply plugs get warm in use, and the repeated heating and cooling tends to loosen the terminal screws. What happens if the earth wire pulls out? At best the plant stops working. At worst you're dead through an unearthed set. Check for screw tightness every few weeks, but don't overdo it as this can lead to localized heating.

3.6

3.6. To be certain of UV eye protection, look for either a CE/DIN mark, BS679 or EN 166 on a welding filter before you buy.

spans the gap between the electrode and work, generating temperatures of around 9,032°F (5,000°C) — which is one reason why you can't arc-weld metals with a low melting point, like lead or zinc; they can't stand the heat. During the weld the rod coating burns, producing gas to shield the liquid weld pool from atmospheric contamination and ionized particles to keep the arc running smoothly. At the same time, flux from the coating floats weld impurities to the surface, where it solidifies as a crust of slag. Molten droplets of core wire are pulled across the arc by magnetic force and surface tension, adding extra metal to the joint. Which is just as well or you couldn't weld uphill, downhill, or overhead.

General points on equipment and procedures are in pictures 3.4-3.19. Do take a very good look through these, as keeping yourself (and others) in one piece is not unimportant. See also the section on safe working ("Pain in Various Forms").

MMA welding heat usually comes from a simple transformer (3.4). AC output models are the cheapest to make as they run straight from an AC power source, like the mains or a generator, and are easily the most common — though DC (direct current) versions are gaining ground. What's the difference? AC welding current varies 50 times every second in a continuous wave, going from positive to negative and back again. In DC operation the current doesn't go through zero, with the result that a DC arc tends to be more stable. But as DC sets need a rectifier to change AC to DC, the sets tend to cost more.

In both cases single- or three-phase mains power is transformed down to around 50V. Most sets offer at least one higher alternative output, with 80V common. In simple sets, welding current is adjusted by a selector which alters the relationship of windings inside the transformer. Increasingly seen these days is the welding inverter — a clever box of tricks that uses electronics plus a small transformer to deliver current. Lightweight and portable, MMA inverters are usually DC sets which, given the necessary torch and gas adapters, can be used for TIG too.

Pain in Various Forms

Arc eye is the nastiest short-term hazard in any electric welding process. It's caused by high-energy ultraviolet radiation from the arc burning the retina, and will happen if the arc is viewed directly or indirectly. Short doses of UV produce a very painful "gritty" feeling in the eyes a while after exposure, with the effects lasting a few hours. Bigger doses can end in blindness. Either way is not to be risked.

Bystanders can suffer arc eye too, either from reflected light or by looking alongside (rather than directly at) the arc. UV light will also burn exposed skin. So it's good practice to make a dedicated welding bay, painting the walls a dark matte color to cut reflections and

If you have a heart monitor or pacemaker fitted, don't weld. The strong magnetic fields around leads and equipment may cause the unit to malfunction.

curtaining it from outside view. When working outside the bay, warn helpers and children about arc eye danger and keep curious livestock away.

A good welding filter is the only protection (3.6). "EW" on a filter glass means it's designed for electric welding. The higher the code number, the denser the filter on a scale from 8 to 14. Use EW 10 or 11 for most work; lower numbers only protect up to currents of 100A, while the highest numbers cut too much light at farm-size welding currents. Keep arc filters scratch-free with a plastic or glass screen, wiping it often during work (3.7) — you can't see much through fog. Be very sure there's no gaps around the mask's filter housing where arc light can creep in. And whatever you do, NEVER use gas-welding goggles for arc work: they don't filter UV light at all.

Which mask to use? The lollipop-on-a-stick sort that usually comes with a cheap welding set doesn't last and often isn't enough for full protection. A flip-down welding helmet lets the operator use both hands when welding or tacking, though these don't suit everyone. If you're planning a lot of work, an automatic self-darkening welding mask is a fine investment: go for one with adjustable shade density and, ideally, manual and grinding modes. Various makes are on offer from welding suppliers.

Electric-shock "tingle" from the rod while changing electrode is unpleasant and reflects the relatively high voltages present. The problem is usually worse in damp weather, when current conduction to earth is better. Wearing gloves is the answer.

Which takes us to **clothing** (3.8). Go for leather boots on the feet, as sneakers or rubber boots offer no resistance to white-hot metal spatter. Back these with cotton trousers minus spark-catching cuffs, then cover everything in non-synthetic overalls or a stockman's jacket, done up at the collar to stop flying sparks and slag. Leather gauntlets protect against ultraviolet light and

3.7. Fume can soon fog up a mask — keep it clean.

3.8. What the well-dressed welder is wearing this autumn.

3.9. Fixed or mobile extractors pull away fume, making them the ideal (if expensive) solution for a welding bench.

3.10. A ventilation pack supplying filtered air to the welding mask is an option if fume can't be removed at source.

3.11. No matter whether you're using fume extraction or not, work out of the rising column of smoke.

stop hot metal disappearing up your sleeves when welding vertically or overhead. Clear goggles look after the eyes during slag chipping or grinding, while a cap hides any bald patch from showering sparks.

Electrode fume is not the best of stuff to breathe, so OSHA legislation sensibly insists on operator protection. Ideally, remove fume at source with a fixed or mobile purpose-made extractor (3.9).

Where fume concentrations are relatively low, a ventilated welding mask (3.10) supplies clean air from a filter pack and simultaneously streams a cooling breeze across the face. Otherwise a disposable respirator designed to fit under a welding helmet may be enough. If taking this route look for EN149 FFP2S as a minimum spec: examples are 3M's 9925 or Moldex's 2485.

Where work is intermittent and good air movement can be achieved by opening doors or windows, fume will be moved on and extra protection may not be needed. If in doubt, the local OSHA office will advise. Whichever arrangement is used, try to work out of the rising fume column (3.11).

Grinding away paint and plating to either side of the weld line minimizes fume problems and maximizes the chances of a good job. Be sure to grind off yellow cadmium plating, for its fume can be a killer. The blue-white smoke from overheated zinc galvanizing is nasty stuff too, at best making you feel ill if breathed for long. Drinking milk may help if you are caught out, but check with your doctor. Watch for hidden plating on the inside of pipes. Sundry paints and degreasants give off poisonous vapors when they get hot, so there's plenty to think about. Take no chances with fume: *ventilate, filter and/or extract.*

So that's a fine list of dos and don'ts. But most are common sense — try picture 3.16 (page 31) as a test.

Self-Tune-Up

In welding (as in life) it's bad to be tense. *Relaxation is central to good control.* There are two schools of thought on how to achieve it — either support yourself so only your welding arm needs to move, or develop freestanding control in a legs-spread position (3.12). The first is easier, the second helps you work when there's nothing to rest on.

Occasional welders prefer the first option. Lean on the bench or machine. Take the welding lead's drag from your welding arm by pinching the cable between hip and bench or draping it round something handy — but before you do, make sure the insulation is 100% complete. A flip-down visor or (better still) an auto-darkening helmet leaves your second hand free to steady the welding arm. Whichever strategy you choose, don't grip the electrode holder tight; it's a sure way to produce wobbles at the rod tip. Try relaxing that iron grip during work and see how control improves!

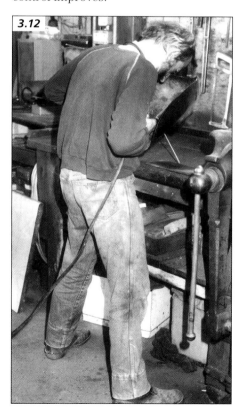

3.12. Stay relaxed for maximum electrode control.

Don't Convert—Invert

Traditional power supplies need big, heavy transformers to handle the substantial welding current flows. Inverters stand this notion on its head.

An inverter takes a 50Hz, 230-240V AC supply, rectifies it to DC, then uses a transistor bank to produce square-wave AC at 8,000Hz-40,000Hz. At these very high frequencies a small transformer can handle the conversion to low voltage/high current welding power, so the physical size and weight of the set can drop dramatically. After transforming, welding-level AC power is rectified back to DC for output, or modified and used as AC. Electronics carry out and control the processes on either side of the transformer, allowing waveforms to be tailored closely to different welding requirements and reacting very rapid to changes.

Which Rod?

There's no shortage of electrodes to choose from, varying in core wire make-up, diameter and flux type.

Core wire must match (or be compatible with) the metals to be joined. If you're not sure on this, ask a welding supplier; your local agricultural dealer may or may not know the real answer.

Mild steel needs a mild steel electrode, fancier stuff needs fancier rods. So be aware that it's a waste of money using a tough, expensive alloy steel electrode to join mild steel in the hope of getting a stronger result. If the rod is matched to the material and your welds are still failing, then your technique is at fault.

Wire diameter. The narrower the rod, the smaller the arc and the smaller the heating footprint it makes on the metal. Thus thin section steels need skinny rods if arc heat isn't to burn straight through them. Fusing thick sections needs more heat; a bigger arc from a thicker rod.

While you're learning, practice with ³⁄₃₂" (2.5mm) diameter rods. Their short length and small weld pool are relatively easy to control. They'll do for one-pass welding up to ⅛" (3mm) and can be used for preliminary runs in heavier stuff. As joints get bigger, move up to rod sizes ⅛" and ⁵⁄₃₂" (3.25mm and 4mm). These three diameters cover the majority of farm work.

Note: When talking about the capability of a rod or set, *thickness* refers to the weld's fillet depth — not the thickness of plate that can be joined!

Flux covering is decided by electrode type and application. Most general-purpose mild steel rods have a flux based on the mineral rutile, with other ingredients to improve arc stability, reduce moisture pick-up in storage and generally make them more sociable to use.

Cellulosic coverings enhance penetration, and are on rods designed for specific purposes like vertical down work and pipe welding.

Low-hydrogen coatings are usually found on rods intended for high-strength alloy steels, possibly also for service at low temperatures. Very sensitive to moisture content, rods with low-hydrogen coating must be baked before use.

Other electrodes add iron powder to their flux, which bulks the volume of weld metal laid down when compared with normal rods.

So which of this lot do you go for? In practice, rod selection for general farm use is pretty straightforward. Find a knowledgeable specialist supplier and buy mild steel rods from a major maker like ESAB, Murex, Oerlikon or Eutectic. Minor brands can be good (and are usually produced by one of the big

Buying an MMA Set

Most farm jobs can be handled with 180A-200A maximum welding current, which is achievable from a single-phase set. Such units allow mobility around the holding and can be powered by a generator, though some inverter power supplies may not like its potentially spiky output — see below. Higher-output welding sets are more bulky, harder to move, and usually need three-phase mains supply, though work well within their capacity at normal welding currents.

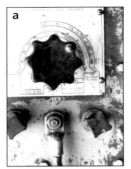

Oil-cooled, copper-wound units are designed for continuous use, running as long as needed on full blast. A typical AC model has 50V and 80V output lines, allowing most types of electrode to be used. (a, left and right terminals). These open circuit voltages (OCVs) are explained on pages 33-34.

Some MMA sets have fan cooling (b). When the windings get too hot a thermal trip shuts the set down, then after a while it comes back to life. The Murex example shown produces DC output which makes for a consistent and very controllable arc, but was more expensive than the many AC alternatives.

The simplest air-cooled sets (c) have no fan, which means a shorter operating period before automatic shutdown and a longer wait before restarting. Consequently they tend to be DIY items. Although such cheaper sets can produce acceptable work, the arc is often not consistent or smooth-running.

A set's **duty cycle** should be checked before buying. It's the time within a ten-minute period that the unit can operate at the current quoted. So a duty cycle of 60% at 150A means

Three alternative sets, each giving different periods of operation. The oil-cooled, copper-wound set (a) is designed for continuous use: it'll run as long as needed on full blast. The AC output model shown has both 50V and 80V output lines (bottom, left and right) giving flexibility over electrode type usable; these are open circuit voltages or OCVs.

The fan-cooled set (b) has vents for forced-air passage. It'll run until its windings get hot, then a thermal trip operates. After a short cool-down period it comes back on stream. This Murex example produces DC output which makes for a smoother-running arc, but it's more expensive than the AC alternative.

Very common on farms is the simple air-cooled set (c), which is quite capable of producing acceptable work. There's no forced-air cooling, which means a shorter operating period before automatic shut-down and a longer wait before restarting.

the set will run for six minutes in ten at 150A. At lower currents it'll run for longer. The higher the duty cycle, the longer work can go on without the set shutting itself off — important where a lot has to be done.

Inverter sets (d) use electronics to do away with heavy transformer windings, so can be very light for their output current. This makes them ideal for site work or machinery field repair. But the circuits in some designs are easily fried by sustained over-voltage or sharp voltage spikes, both of which generators can produce. As a rule of thumb, an inverter should only be paired with a generator whose capacity is at least 1.5 times the set's power requirement. For example, a set drawing 5kVA when delivering maximum welding current should only be used with a generator able to

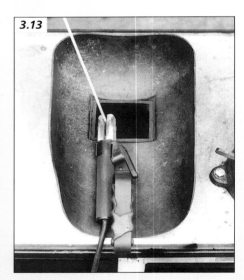

3.13. A good habit to cultivate. Laying the electrode holder in your mask or on something insulating (wood is good) means you know where it is (!) and lowers the chance of accidental arcing. Give gas cylinders a wide berth with the electrode, as accidental contact and the resulting arc could be terminal. It follows that chaining gas cylinders to a metal welding bench is not a bright idea.

deliver *at least* 7.5kVA. If in doubt, check this out with the supplier — better safe than sorry.

An inverter's electronic control lets the maker add sundry operator-friendly features. These can include:

- **Hot starting.** To help the arc to start and to boost initial fusion, welding current is increased automatically for a short while as the arc is struck.

- **Anti-stick.** If the rod sticks to the work, the set senses the sudden sustained rise in current draw and cuts off after two or three seconds. Saves a lot of agitated hopping about.

- **Arc force.** The less experienced (or more tired) the operator, the more arc length varies during a run. The power supply spots the rise and fall in current, and within limits compensates for them. Input current is thus steadier, minimizing changes in fusion and penetration and boosting weld quality.

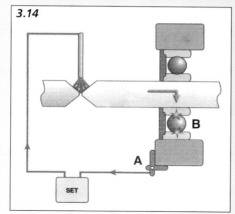

3.14. Small things can turn out expensive. Clipping the return lead to a bearing housing (A) forces current through ball or roller contact patches on its journey round the circuit (B). If these burn and/or develop flat spots, the bearing turns to junk. The thoughtful operator also works out how far gas or weld heat is likely to travel: bearing seals are easily cooked.

This "active" compensation goes beyond what's provided by a normal, non-electronic MMA power supply — see pages 51-52.

3.15. Early alternators could be damaged by welding's relatively high voltages, though modern versions have inbuilt protection. If in doubt, disconnect the battery and alternator before welding on a machine or anything attached to it. Where there are electronic control systems (modern tractors, combines, self-propelled foragers and so on) disconnect the battery and ideally take out any plug-in modules. Alternatively connect a spike catcher (inset) across the battery to damp out surges. Automotive factors stock them.

3.16. How many potential nasties can you spot here? Answer at foot of page 33.

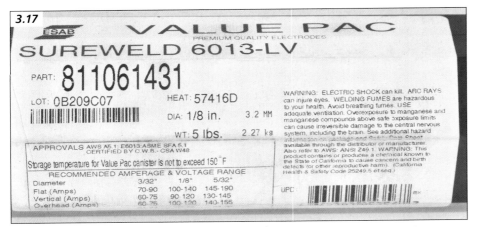

3.17. *Rods from major makers carry full information on the packet. Listed are the material(s) that the electrode can weld, positions it can be used in, suggested welding current range and the required OCV. Preferred polarity is important with DC sets; that is, whether the rod should be run positive or negative.*

makers anyway), but starting with a big-name brand is always a sound idea. The rod packet should list all the info needed to work with the rod (3.17).

Stock up on sizes ³⁄₃₂", ⅛" and ⁵⁄₃₂" (2.5mm, 3.25mm and 4.00mm), adding a pack of ¹⁄₁₆" (1.6mm) "sparklers" if sheet steel work is intended and you don't have gas or MIG/MAG — but these very small rods can be unexpectedly expensive. If vertical down welding is to be done, make sure that the rod flux produces fast freezing slag — see page 45.

For welding anything other than mild steel, know your set's OCV capability before talking rods with a welding supplier. Firms usually give advice freely, along with lists of electrodes and their applications. As mentioned earlier, "dissimilar steels" rods cover most agricultural eventualities. For instance, Oerlikon's Inox DW will weld steels of varying carbon content, low alloy and spring steels, ferritic and austenitic stainless steels to themselves . . . and to each other. And as dissimilar steels rods are relatively hard, most can be used as a "buffer" layer between mild steel and a hard-facing layer, reducing the risk of dilution which can cause the expensive topcoat to crack and flake away.

Buy specialist rods only when required, as they're expensive. Low-hydrogen

electrodes are a hassle. In non-critical work, dissimilar steels rods will probably cope anyway and are far less trouble to store and use. Cast iron is a law unto itself and needs dedicated rods and technique; see page 135. Building up worn parts or laying hard metal on new soil-engaging bits calls for specialized hard-facing consumables. Tell your rod supplier the application and he'll recommend something appropriate.

Watch Out

Time to say it again: the cardinal rule is "be cautious." *If in doubt over rod selection or technique, let a specialist do the work.* That may more expensive . . . but then DIY can sometimes end up very costly in unexpected and unpleasant ways.

Finally, a word on rod storage. Keeping rods in their original packet means you hang on to the information. And although good-quality mild steel rods can survive all year round in the workshop if they must, eventually the damp will get to the coating. Best to keep them dry, for which an old fridge modified so the bulb always burns is good. Then at least you'll know the answer to the old question: what happens to the light when you shut the fridge door?

Build Good Technique 1: Getting Started

Until you get the knack, striking and holding an arc can have you tearing out hair. But there's a simple way — scratch, tap and lift.

With the right current dialed in on the set, SCRATCH the rod tip on a piece of earthed scrap to get it hot. Current flows easier from a warm tip. Be positive with the scratching; use a wide sweep that generates showers of sparks.

Bring the rod quickly into position over the weld area, and in one smooth movement TAP it down and LIFT it just a little (3.18). The arc will establish during the lift and you can weld on.

Beginners often are understandably tentative, touching (rather than tapping) the rod on the plate at too shallow of an angle. Come down firmly at a near-vertical angle to get the arc up and running. But if the rod sticks to the work, break contact quickly by circling your hand and pulling up.

Sometimes the rod wants to stick even though you're doing all the right things. Check its tip (pictures 3.20a, b and c). If this is healthy, look next at the return lead and clamp (3.21). The contact patch between clamp and work must be clean for easy current flow, so grind a bright area on the work and attach your clamp there. If the return lead can't attach directly to the bits being welded and the

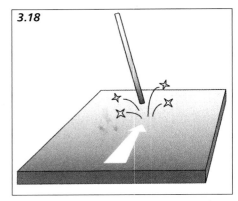

3.18. *Getting the arc started is easier if you warm the tip by scratching it on a plate. A generous sweep (arrow) produces the best effect.*

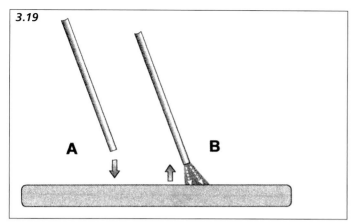

3.19. To start an arc cleanly and quickly, tap and lift the rod in one positive movement.

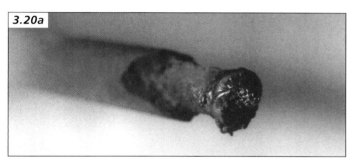

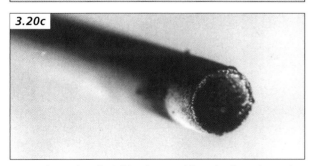

3.20a, b, and c. If sticking is a problem, look at the rod tip first. Is it like 3.20a or 3.20b? Condition 3.20a comes from heavy-handedness with a sticking rod or damp flux. To rectify, scratch the rod hard until the bare core wire is melted away. Condition 3.20b comes from a carelessly broken arc at the end of a previous weld; the slag has sealed over the core wire. Scratch and/or tap it hard to break the seal. Condition 3.20c is good; bare core wire waits inside a cup of flux, ready to establish an arc.

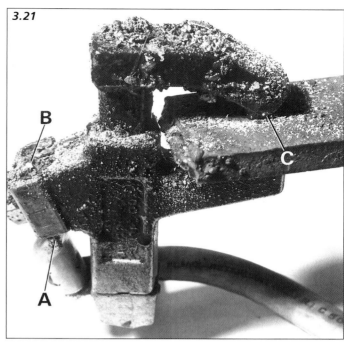

3.21. Free current flow is choked by a return lead clamp that's past its sell-by date. Typical trouble spots are the thinned or frayed cable (A) and corrosion where cable and clamp come together (B). Even when the clamp is healthy, big trouble comes from poor contact between the clamp and the job (C).

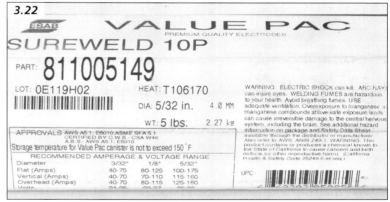

3.22. Still no arc? See that there's enough current coming from the set. Here the pack suggests 50A-90A for a ⅛" (3.25mm) rod, so 80A should see easy starting. The required OCV (bottom left) is 50V, a common value delivered by most sets.

Answer to 3.16: Think sparks to see what can happen. A battery on charge gives off hydrogen gas, which can explode on contact with hot spatter. The rubber gloves and cloth at the back are likely to catch fire; the electrode holder is not in the mask and its rod tip lays perilously close to the work, risking an accidental arc. Cans of petrol, penetrating fluid and engine oil sit alongside the welding area, while the upended chainsaw will leak vapor or neat fuel. At floor level the open waste container, cleaning rag box and grease bucket don't bode well. Neither do the inflammable degreasant and oil-soaked sawdust.

current has to go via a welding bench, vice or anvil, make certain there's good contact between the work and whatever it's resting on. A small arc and burn spot often appears at this contact point, so think about that if the job's surface finish or appearance matter.

For maximum current flow, all cables must be in good order with no current-blocking "bird's nest" of frayed connections (3.21, a and b). New return leads and clamps are cheap, so throw out anything that's substandard. Don't use the return clamp as a striking-up area; a piece of scrap is a much better bet.

If the arc still won't start, check that there's enough current dialed in on the set. But how much is enough? The rod pack gives a suggested range — aim toward the top of it for easy starting (3.22). The pack should also show the minimum open circuit voltage (OCV) needed. If the set can't deliver this, striking will be much more difficult.

Build Good Technique 2: The Four Variables

Understanding the four variables is the key to a good job. What are they? Rod angle, speed of travel, and arc length are all under operator control during a weld. The fourth is current.

Get each one right and a strong job results. Get one or more wrong and faults appear: lack of fusion, poor penetration, slag traps, undercut. All things that make welds weak.

CURRENT

Current has the edge in importance. If you're short here, the terrible twins — penetration and fusion — will be lacking. Both are vital for a sound weld. Low current also makes it harder to establish an arc.

The first rule of current selection is not to believe the graduations on the welding set! You'll soon learn to see, hear and feel how the rod is running; to monitor what's happening at the rod tip and to alter current accordingly. Changes in welding position usually call

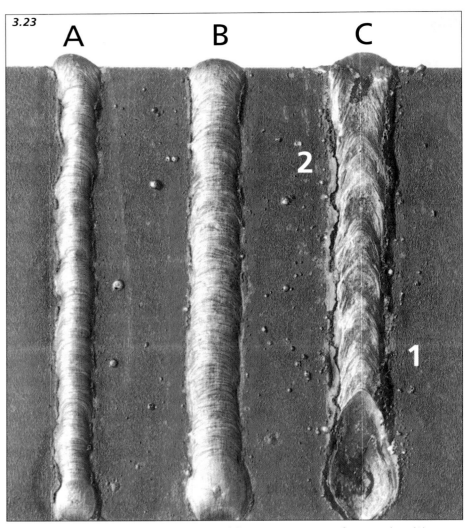

3.23. Fusion and penetration change as welding current goes up from too low (A) to too high (C). Setting (B) was just right. Undercut and trapped slag come with too much heat (C, 1 and 2).

for current adjustment too, which will be covered later.

Use the rod pack's information to set current for a start, choosing a setting toward the upper end of the recommended range. More current means more heat, and more heat generally means easier striking, better fusion and more penetration. If the rod packet information is missing, allow 40 amps per mm rod diameter. On all but the highest-quality work, a little too much current is better than not enough.

To get a feel for current setting, experiment on some scrap. Make short welds, starting from below the

packet-recommended minimum and ending above maximum. See how striking, maintaining the arc and fusion all improve as current goes up?

This is summed in picture 3.23. If there's not enough current (heat), weld metal just heaps on top of the underlying plate (A), leaving fusion and penetration poor. Where current is too high (C) the weld bead is wide and flat, the final crater is deep, excessive heat melts a wide track and beads of spatter go everywhere. The bead surface is also often V-shaped, thanks to the operator speeding up to keep fusion or burn-though under control. Where current is too high (C1) undercut may turn up

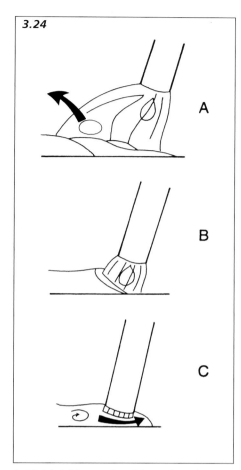

3.24. Variable number 2 is arc length. When it's too long, filler metal will skip out as spatter (A). When it's too short, slag can run round the tip to for traps (C). When it's spot-on your eyes and ears will tell you.

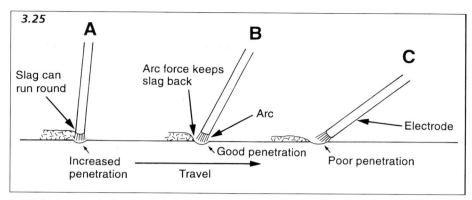

3.25. Rod angle during work has a big effect on fusion, penetration and slag trap production. A and C lead to problems, whereas like Goldilocks' porridge, B is just right.

where the bead and plate meet. This happens because as the arc passes, it melts a crater too wide for the rod to fill; and if the arc is simultaneously too long, some of the rod's filler metal will also be lost as spatter. Watch out for undercut, as it provides a notch from which failure can start. Working too hot also traps slag at the weld edge and in the coarse weld bead's surface, making it hard to chip off (3.23, C2).

Where current is about right (3.23, B) weld metal fuses and penetrates into the parent plate — exactly what's needed for a good weld. The bead is consistent in width and the surface ripples are U-shaped; there's no undercut and little spatter.

For most jobs a little too much current is better than not enough. But don't go overboard or you'll waste rod, create undercut, risk burn-through and porosity, cover the job in spatter and increase distortion by feeding in excess heat. And as an encore the rod's flux coating will break down.

ARC LENGTH

Variable 2 is **arc length**. A rough guide keeps the arc just shorter than the core wire diameter, that is, about 0.08" (2mm) for a ³⁄₃₂" (2.5mm) rod. But this is not a fixed relationship.

Keep your ears and eyes open during work. A good arc has a clear, bright area under it, bordered by a semi-circular curtain of molten slag behind (3.24A-C). You'll know when arc length is right, for its noise hardens from a soft, fluttery sound (arc too long) to a crisp crackle as the right length is found. A good-length arc sounds much like the sizzle of breakfast frying in the pan. When the arc's too short the rod dips in the molten slag and everything goes quiet, arc light dims and slag wants to run round in front of the tip. This is a perfect scenario for producing slag traps (page 44).

As your confidence grows so arc length can be used to control slag. If you see a violent swirl starting to develop in the slag just behind the rod tip, a trap is imminent: lengthening the arc a little can save the day. But be careful, for an arc that's too long can itself produce a

trap: losing filler metal as spatter means the weld pool isn't properly filled and slag can take its place. This is a common problem in fillet joints. Experimenting with arc length (while keeping the other three variables right) is the best way to get a handle on what's happening.

ROD ANGLE

Variable number 3 is **rod angle**, which affects both fusion and penetration (3.25A-C). Too shallow an angle lets the arc push weld metal along the work's surface, so there's not much chance of good fusion or penetration. But too steep an angle risks molten slag running round the rod tip to interfere with the arc, causing slag traps (A). The right rod angle (B) steers between these possibilities.

For work on the flat, an angle of around 70° is right. Some electrode holders come with angled grooves in the jaws. If you put the rod in these and keep the holder's handle parallel with the work, then electrode angle won't be far out for down-hand jobs — those in a flat position. It's OK to bend the rod to suit the job — but not so far that the flux coat fractures.

When welding a joint it's usually best to keep the electrode centered laterally over it. Angling the rod toward one plate or the other puts more heat into that plate, which is generally bad for balanced fusion. But deliberately aiming the rod to control heat input can be useful, as when welding thin sections to thick ones — see page 40.

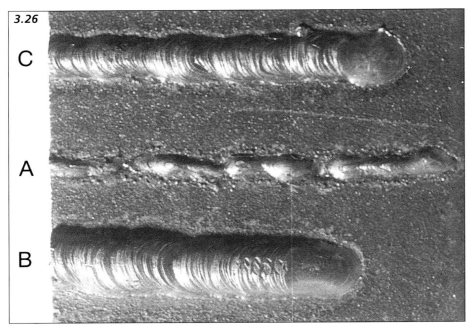

3.26. Going too fast (A) gives no time for fusion or penetration. Travelling too slowly lays down too much metal and the extra heat adds to distortion. Travelling at the right speed leaves a bead that's about twice as wide as the rod's core wire.

Some rods tolerate operation outside normal limits better than others. So experiment with the rods you use, looking and listening as the angle is changed.

SPEED OF TRAVEL

The final variable is **speed of travel**. How fast is fast enough? Picture 3.26 shows the differences. Hacking along the joint at a rate of knots (A) leaves no time for weld metal to be deposited, so the bead is stringy and very weak. Travelling too slowly (B) takes longer, wastes rod, increases distortion by increasing heat input and leaves a large humped deposit. But moving along at the right speed (C) leaves a neat, even bead roughly twice as wide as the electrode core wire: that is, a 5⁄32"-6⁄32" (4mm-5mm) deposit from a 3⁄32" (2.5mm) rod.

Varying your welding speed during a run is bad, too. Look closely at 3.26A and you'll see that even in this acceptable weld, the bead necks down in places along its length. This shows that the speed of travel was not constant. In the fast sections the weld changed thickness, varying joint strength; surface ripples pulled into V-shapes are a giveaway. When welding try to imaging you're a machine, geared always to travel at a set speed down the joint.

So there they are, the four variables. Generally, first-time MMA operators use too little current, work too fast with too-long an arc and tend to let rod angle drop during a run so it ends up too shallow.

Making the Weld 1: Keep It Clean

Now to join some metal. Step one is *clean before you weld*. Never mind italics: write it in letters of fire. Agriculture produces a fine range of contaminants ranging from rust and paint to oil and cow dung. None of these need to kill the arc before they compromise joint strength: the inclusions or weakening pinholes they create are enough. So sort out contamination before you begin, no matter what the electrode salesman says.

And if preparation isn't thorough? Rust on the weld area will turn into brittle, weak iron oxide particles in the joint.

So grind off (or flame clean/wire brush away) all rust, over and around the weld area.

Paint, oil and animal outfall all vaporize under the arc, leaving blowholes in the weld varying in size from the invisible to the substantial. Normally a run round with the angle grinder is enough to clear crud. But on jobs where oil or other stuff has really penetrated (like an old oil-containing casting) it's often necessary to make a first run to draw out contamination, then grind that away and weld again.

Whatever the contamination, clean an area around 2" (50mm) wide on either side of the joint. That might seem a lot, but it's the best way to minimize fuming from paint or plating. Grind galvanized steel until it sparks bright, showing that all the plating has gone. Joints should get a coat of paint afterwards, so all this generous preparation won't be wasted.

We've already mentioned metal platings but they're serious enough to re-run the subject. Look out for cadmium plating's dull yellow finish and grind all traces of it away, as heavy metal fumes are really nasty. The zinc in galvanizing makes a mess of a weld, and the dense blue-white smoke produced does you no good at all. Where there's a risk of fuming, invoke nature's free ventilation by working in moving air. Indoors, use a fume extractor or the appropriate respirator. Your welding supplier should be able to set you up with the right protective equipment.

Making the Weld 2: Get Set Up

Now look to preparing the joint metal itself. Overstating the importance of this stage is hard, but still it's often skimped. Think on the following for a moment:

Any joint must be welded to full depth to duplicate the original material's strength.

So if you're talking butt joints (i.e., two plates laid side by side to be joined by their edges), welding to full depth almost always involves some sort of edge preparation. That is, cutting or grinding away the edges to let the weld get to the bottom edge of both plates.

Material thickness is the trigger. Butts in mild steel up to ⅛" (3mm) thick can be welded without preparation, but thicker stuff — which means with most farm job, whether fabrication from scratch or a simple repair — needs edge preparation.

Just leaving a gap between otherwise-unprepared edges does help penetration,

but not much. Despite all the noise and fury, an arc does not dig very deep. Check this for yourself by welding a series of butt joints in ³⁄₁₆"-¼" (5mm-6mm) plate, leaving an increasingly large gap between the plates each time. Weld one side of the joint only, and then break it open in the vice and look for penetration. You'll probably conclude that the arc needs help to penetrate. The best way forward is to bevel the plate edges with a grinder or gas cutter.

Illustration 3.27 lays out the general idea. A total angle of 60° (that is, 30° on each plate) is the best compromise between accessibility and gap size, but in practice grind a little more to allow yourself plenty of elbow room. A wider Vee makes with the first (or "root") run easier: if the angle is too tight it can be a slag trap generator. Dodgy technique (insufficient current, travelling too fast, wrong rod angle or too short or too long an arc) is magnified when welding in a Vee.

Making the Weld 3: Different Joint Types

Here's a run-through of things to think about when MMA-welding various joints.

First, the basic **butt weld** in prepared-edge plate (3.27, 3.28 and 3.29). Try really hard to get good fusion and penetration in the root run (3.27). If the Vee is tighter than 60°, open it out before you start welding. If this can't be done, dial in a little more current than usual to minimize the chance of slag traps. Where the job allows, boost penetration by leaving a gap between the plates' bottom edges, though don't go wider than electrode thickness.

Fire in the first run with a small electrode; ³⁄₃₂" (2.5mm) is usually fine. Using a thick rod at this stage is usually counter-productive as it can't reach into a Vee bottom as well as a slimmer one. Having said that, a thicker rod's extra metal and current can help stop slag traps. Try it and see, but watch for lack of penetration.

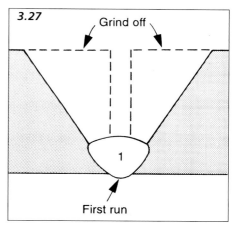

3.27. To get to the bottom of a butt joint in material over ⅛" (3mm) thick, first get rid of the edges with a grinder or gas cutter (dotted lines). Then fire in the first run with good penetration (1).

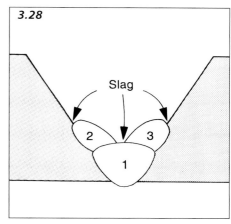

3.28. Fill the Vee using overlapping runs — 3.30 has the technique.

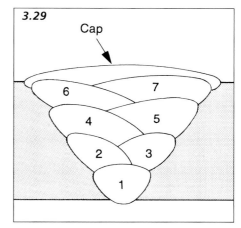

3.29. Finish a wide Vee with a weaved capping run: 3.37 shows ways to weave.

Lower Limit

As material drops to ⅛" (3mm) or below, blowing holes in the work gets increasingly likely. For most mortals, ⅛" (3mm) sheet is MMA's lower working limit.

3.30

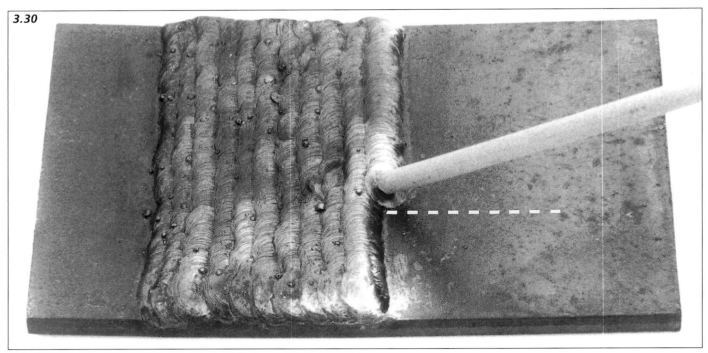

3.30. Overlap technique. Start with a single run, then build a pad by adding more runs so each half-overlaps its predecessor. Angle the rod at 45° and fire the arc at the spot where weld and plate meet. This basic method is brilliant for building up worn parts or re-surfacing.

3.31

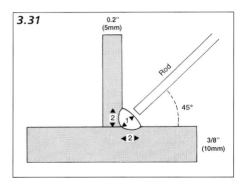

3.31. In T-joints or fillets, weld throat depth (1) needs to at least equal the thinnest plate. Equal leg lengths (2) give balanced strength, and come from a 45° rod angle.

3.32

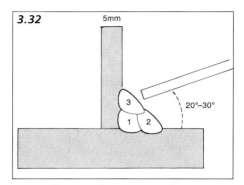

3.32. Often you'll need more than one run to achieve enough throat depth. Sequence these as shown, using a shallower rod angle for the final pass (3) to keep it pushed well up.

With a good root run in place, fill the Vee using the basic overlap technique in 3.30. Place each weld so it either overlaps or stays well clear of its neighbor, as the last thing you want is a narrow, slag-trapping gap between adjacent runs (3.28). Chip off all slag and exercise the wire brush after each pass — good "weld hygiene" leaves no bits of slag behind to form weakening inclusions in the next run.

Joint width grows as you fill it, allowing the use of thicker rods. Switch to a bigger diameter rather than going more slowly with a small rod, or slag traps will result where there's not enough filler metal. If appropriate, finish off the joint with a weaved capping run. This will blend the weld's edges into the original plates and "normalizes" the preceding passes. If you didn't manage to achieve full penetration from the first run, shame on you — but you won't be the first or the last to do it. In such cases, turn the work over, grind the joint line into a U-shape then seal it with a single pass.

To form a strong **T-joint or fillet**, the weld's throat depth needs to be equal to the thinnest plate (3.31). So, when joining 0.2" to ⅜" (5mm to 10mm) plate, the throat should be at least 0.2" (5mm) deep. Equal leg lengths give balanced strength: achieve them by angling the rod so it splits the joint in two, that is, forms a 45° angle.

Good fillet root fusion comes from a mix of careful plate alignment and steady work with a small diameter rod. If plates aren't parallel, one inevitable takes more heat than the other and the gap between them varies. And it's easy to make a slag trap at the start: the best way around this is to let weld metal fully fill the joint before moving off, keeping speed down if necessary to be sure it does. Once fusion spreads across the joint, move off. Keep travel speed steady and arc length relatively short to make certain the joint is filled and free from slag traps.

Thick plates will need more than one pass to get the required throat depth. In

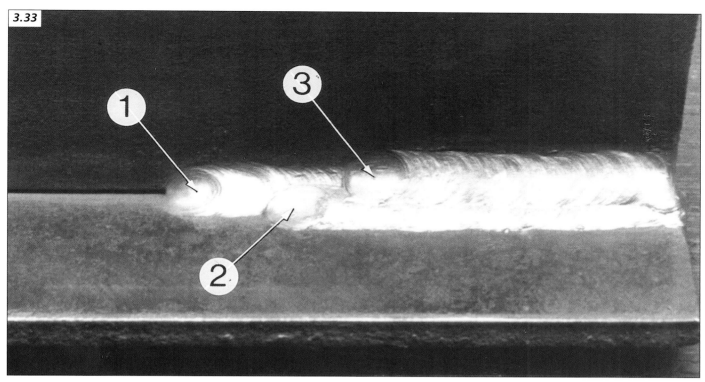

3.33. The theory of 3.32 turned into metal. Starting off slowly let the three overlapping runs fuse full-width at the beginning of the joint (far right). See how they then blend into each other and the parent plates?

a three-run joint, sequence the passes as shown in 3.32: see how number 2 gives number 3 a "shelf" to sit on? Keep fresh weld metal well up on the vertical plate during pass 3 by using a shallower rod angle. And where necessary, go up one rod diameter for runs 2 and 3 to fill the space.

If slag traps are a problem, check the four variables. Look at current first: too little is a prime cause. Welding

Run Number Alert

Don't go overboard when making a multi-pass fillet. Use the minimum number of runs needed to build the right throat depth. Why? Heavy weld build-up often makes a joint weaker rather than stronger, as heat build-up can affect material properties and stress is concentrated at the edges of the weld.

with limited amps (=limited heat) is always dodgy, as the bottom plate in a T-joint acts a big sink, draining heat from the weld. On fillets use 10%-15% more current than for the equivalent butt joint, but don't overdo it: check the vertical plate for undercut at the bead edge, as significant notching produces a weak area.

Where current setting looks good but pesky traps are still appearing, think about your speed of travel. Going too fast doesn't give time for the joint to fill, inviting in slag and leaving undercut. Arc length also matters. If it's too long, metal takes a short cut and goes to the joint walls rather than the root. If the arc is too short, slag can run round the rod tip and mess with the arc.

After this, **lap joints** are relatively easy (3.34). Tack plates together on the non-welded side where possible, and truck along at a speed such that the upper plate's top edge just melts into the weld pool (arrow). If the joint looks "hollow" or won't fill properly,

try the next-fatter rod. As with T-joints, slag traps come from too little current, travelling too fast or the wrong arc length.

Open corner joints — ones where plates meet to make a "roof" and are welded on the outside — are just as straightforward as laps. Take great care over set-up; this is half the battle. Tack the joint halves so their edges are parallel, leaving a small penetration gap in material over ⅛" (3mm) thick. Travel along the joint at a speed that sees both outer plate edges melting into the weld pool. Use multi-runs to fill the Vee in thick material, sequencing passes as in 3.32.

Turning over an open corner joint leaves a trough. For practice, weld this just as you would fill a prepared butt, using overlapping runs to build thickness as required. Deep Vees attract slag traps like a dead rabbit attracts flies, so use a little more current than normal to cut the risk.

3.34

3.34. Start a lap joint by tacking it so the plates are in good contact. Any gap risks the top plate's exposed edge burning back. Travel at a speed that just melts in the upper plate edge (arrow).

Horizontal-vertical joints are those that run horizontally between vertical plates, as 3.35 shows. Welds in this position can end up with globs of weld metal to rival Babylon's Hanging Gardens. Too much heat — from high current setting or travelling too slowly — is the usual cause, made worse where a long arc lets weld metal drop out of the joint. Dial in only enough current to give penetration and a smooth-running arc. Angle the rod to keep the bead pushed up into the joint, keep a constant and shortish arc. Don't travel too slowly.

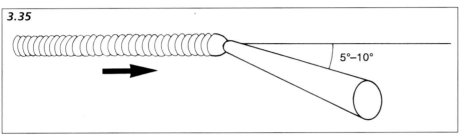

3.35

5°–10°

3.35. To keep a horizontal-vertical joint tidy, use a short arc, minimum current and don't hang about. Angle the rod as shown to keep the bead pushed up into the joint.

Last but never least, welding **thin sheet to thick plate** is not as hard as it might seem if the two are kept in close contact. Picture 3.36 shows ⅟₁₆" (1.6mm) sheet lap-jointed to ¼" (6mm) plate. The corners were tacked to hold the plates tight together, then each side was welded. The arc was aimed so that most of its heat went into the thick plate, leaving just enough warmth to fuse in

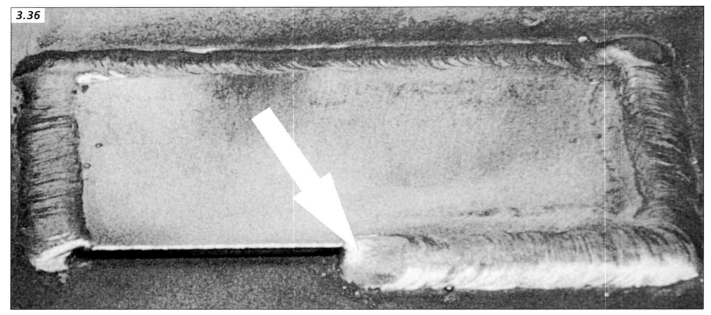

3.36

3.36. Welding thin sheet to thick plate is easy as long as the pair stay in close contact. Fire the arc mainly into the bottom plate, letting arc heat just melt sheet metal into the bead.

the thin sheet's edge (arrow). Watch the sheet during operations; if it starts to buckle, it'll burn back quickly and you'll end up in a pickle. Stop, dress the sheet back down and continue.

Making the Weld 4: Stops, Starts and Other Business

Here are several techniques that will make the finished job stronger and neater . . . or get you out of trouble. The principles apply to MIG/MAG work too.

Weaving is a useful trick used to spread heat and metal. You'll need to weave in all sorts of places: making a "capping" run on a substantial multi-pass joint, when putting down a layer of hard-facing metal, when welding uphill or downhill, when bridging gaps between plates. Whatever the job, the important thing is to follow a definite pattern and work very deliberately. Pictures 3.37 and 3.38 show how.

Whatever the purpose, weaving is best kept simple. First because that's easier (never a bad thing) and second to lessen the chance of trapping slag in the weld. Either a crescent weave or Z-shaped zig-zag serves most purposes other than vertical work.

How wide should you go? A rough rule of thumb allows up to three times electrode diameter, otherwise the weld's edges may cool too much and you'll get slag traps and/or "cold shuts." In a cold shut, fresh metal lies over (rather than fuses with) the stuff you've just laid down, forming a weak point. The motto is "Take care of the edges and the middle looks after itself," which means making a deliberate pause at each side to let metal flow, fuse and build before moving smartly across the weld face.

When you're weaving, make very deliberate rod tip movements. Waving the rod about pushes heat and metal around, inviting slag traps. Keep those at bay by weaving tight into the face of the weld, that is, not travelling too far forward for each traverse. On jobs where a conventional weave would be overstretched, infill weaving between

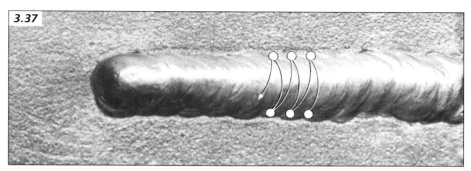

3.37. Either the crescent weave (as here) or a zigzag pattern is fine for capping runs. Be very sure to pause significantly at each edge (white dots), letting weld metal fill and fuse. Slag traps are likely if you rush or wave the rod tip about.

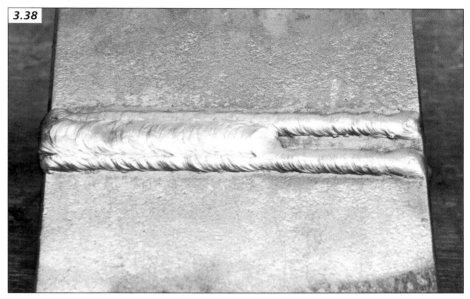

3.38. Where a simple weave would be overstretched, filling between two beads is the answer.

two beads can help (3.38). Patterns for vertical work are covered on pages 44–46.

First cousin to weaving is **building up a thickness** of weld metal. You'll do it while filling a deep prepared-edge joint, when remaking worn bits or when hardfacing soil-engaging parts. Picture 3.30 (page 38) has the basics. First lay down a single bead, using a chalked guide line if it helps. Then add successive runs so that each overlaps the one before. Use a 45° rod angle, firing the arc right into the place where the last run meets the plate. Adjust speed of travel so the new metal half-covers the old. Chip off, wire brush and inspect after each run. Repeat until the

thickness needed is achieved, watching out for distortion through heat build-up.

A small snag with MMA is that you have to keep changing rods. How you manage **stops, starts and restarts** has a big impact on weld strength and appearance.

Let's start with **stops**. As a rod burns back keep an eye on the flux coat, particularly when using high currents. If it starts to fly away from the tip in beads, stop the weld right there; protection from the coating is disappearing. Plan on finishing all runs anyway when the rod burns to a 2" (50mm) stub. It's never worth trying to get to the end of a joint before the rod expires: you seldom

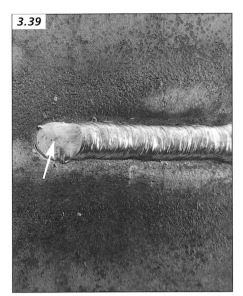

3.39. To set things up for a clean re-start, break the arc cleanly as the old rod burns back to 2" (50mm) length. This will leave a hollow pad (arrow).

make it, and just end up leaving a poor stretch of weld.

As the old rod comes to its end, set things up for a good restart by breaking the arc cleanly. Whisk the tip away to one side, leaving a shallow crater (3.39). Wire brush the area clean.

To **restart** with a fresh rod, warm the tip by scratching it or make a short run on some scrap — not on the return clamp! Quickly position the tip about ½" (12mm) forward of the waiting crater (point X in 3.40), tap it to start the arc and immediately lift the rod. Lifting does two things: gives enough light to see where you're going and minimizes the metal laid down. Keeping the arc relatively long, smartly move the rod tip back to the crater's forward edge (3.40, dotted line). Close the arc down to welding length, then make a very deliberate "wall of death" passage around the crater to fuse old and new metal (3.40, solid line). Don't rush! Move slowly forward out of the crater, picking up normal welding speed as you leave. Done well, you won't be able to feel the restart with a fingertip and will have trouble seeing it (3.42).

3.40. To restart, tap the warmed rod tip at point X. Keeping a long arc, move it to the crater's front edge (dotted line), then close the arc to welding length and make a very deliberate pass round the crater. As you come back to the entry point, move on down the joint line.

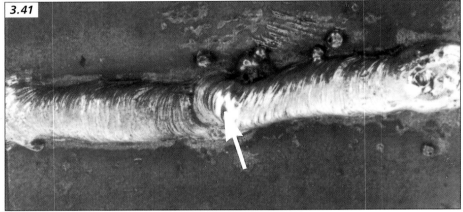

3.41. Waving the rod tip around or rushing during a restart produces a mess (arrow) . . .

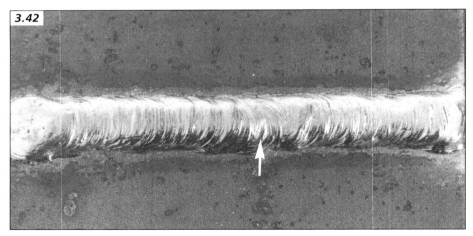

3.42. . . . but if you're slow and deliberate, the restart will be hard to see and will not snag a fingernail run over the area (arrow).

Starts at the beginning of a joint can be pesky. Common is a "snail trail" of weld and a slag trap before a decent weld establishes (3.43). Initially the rod and plates are relatively cold, so metal is slow to fuse. Moving off too briskly (or with a long arc) produces the fault. The trick is to start the arc just off, rather than directly on, the joint edge. Pause a while before moving off, watching to see that molten metal fuses with both plates. A small sideways weave usually helps: try it and see. Where an ultra-solid start is important, use a run-on plate.

3.43. Recognize this? A slag trap at the beginning of a weld (arrow) springs from two things: cold plates and an arc that's too long. Pause before moving off or use a run-on plate.

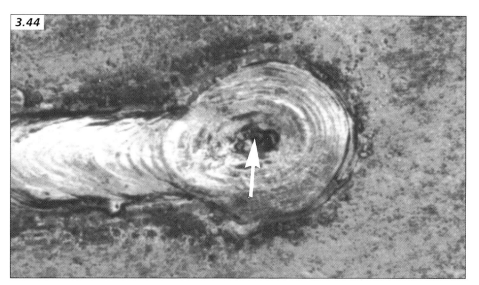

3.44. This nasty finish comes from a combination of bad technique and too much current. The hollow crater (arrow) will be a launchpad for cracks.

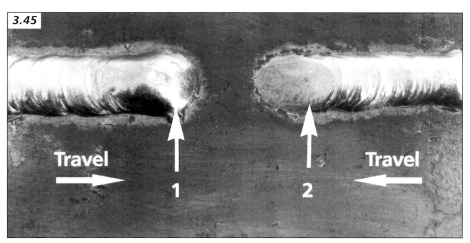

3.45. Two in one here. A flat finish is needed for clean restarts (1). When a joint is closed, the crater must be filled (2).

That is, establish the arc on a piece of scrap alongside the joint then move into the work.

End-of-joint **finishes** matter too. An untidy, hollow finale is a launchpad for cracks (3.44). Although you need a flat stop pad to pave the way for a clean restart (3.45, 2), at the end of a joint the crater must be filled to maintain bead strength (3.45, 1). Producing a good finish takes slow, deliberate action. Rush and you'll spoil it. Weld toward the end of the joint as normal. Then move back a little from the end and pause, keeping arc length constant to pile metal into the crater. Circle the rod tip slightly but not too wide, or weld metal spreads too much. Finish the job by travelling forward a little and breaking the arc with a quick sideways flick of the rod. Concentrate on keeping arc length steady throughout. Don't pull the rod upwards to break the arc, as this may leave a crack-propagating hole in the finish area.

Tacks are the daily bread of welding. But if a tack weld is to lock plates into the right position and hold them there under contraction stress or while things are re-aligned with a big hammer, it needs to be a quality mini-weld rather than a quick blob. Worthwhile tacks in thick material are ⅜"-¹⁵⁄₃₂" (10mm-12mm) long and fused properly into the joint material (3.46). Where you can, tack on the reverse side of a joint to leave the weld area uncluttered. If this can't be done and you're after a clean surface appearance, grind away the tack's bulk before welding the joint.

On long work, tack as often as you need to hold parts in alignment. Don't skimp on tack numbers, as it takes less time to weld them than to put things right after joint components have bowed or shifted through lack of anchoring. Two hints for quality tacking:

• Save part-used ³⁄₃₂" (2.5mm) rods for the job. A short rod is easier to control, and its small diameter means good penetration in a Vee.

• Use more current than normal to give an easy start and quick fusion. Either shift down one rod size (e.g., from ⅛" to ³⁄₃₂" [3.25mm to 2.5mm]) while leaving current setting unchanged, or use a ³⁄₃₂" (2.5mm) rod on ⅛" (3.25mm) settings.

Slag traps are a fact of life. Good preparation, sound basic technique and lots of practice keeps the little burglars at bay — but what can you do when they show up? There are two options. Either grind out the joint and reweld it (which won't be possible if you can't get a grinder into the area), or use arc heat to melt out the slag and re-fuse the trap area (which is not always 100% successful).

Where possible go for option one. But the second approach will do in a crisis. First get out all the slag you can, using anything pointy. Rewelding aims to float the remaining slag to the surface and fuse fresh metal into the joint, so violent measures are called for. Put mega-heat into a limited area by using a lot of current on a small electrode — say 130A on a ³⁄₃₂" (2.5mm) rod. Then strike up just short of the trap, get the arc running nicely and move slowly over the dodgy area, watching to see that the arc digs out all the crud and the void is filled with fresh metal. Picture 3.47 has gory details.

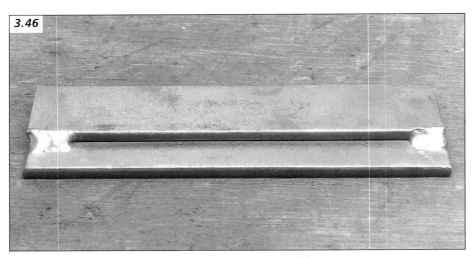

3.46. Tacks need to be good-quality "mini-welds," not blobs.

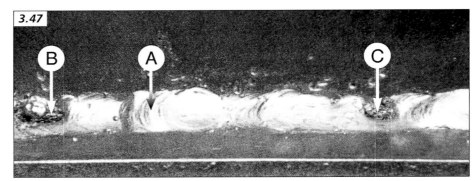

3.47. If you can't grind out slag traps, use a small diameter rod and high current to re-fuse them. Here A has been given the treatment while horrors B and C wait their turn.

Making the Weld 5: Up, Down and Overhead

Now for the bit that brings some operators out into a cold sweat — working straight up, straight down or overhead. If there's no chance of taking the chicken's choice — rotating a vertical job so it's flat — you'll have to decide whether to weld uphill or down.

Vertical down work is usually favorite as most people find this easiest. But penetration, and hence joint strength, will be very poor in all but thin sheet. Welding vertically down is only good for sheet, for cheering up surface appearance, or for filling awkward gaps where burn-through would otherwise be a problem. *Never rely on it where strength matters.*

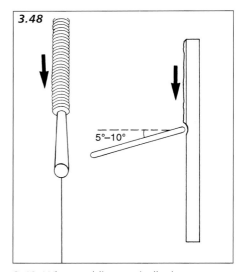

3.48. When welding vertically down, center the rod over the joint to put heat equally into both plates. Angle it just below the horizontal and keep arc length shortish, then travel at a speed that keeps the rod tip just below the descending slag curtain.

Success hangs on having the right rod and enough current. Fast-freezing slag is called for, which will stay above the arc rather than running down under it and getting in the weld pool. Oerlikon's Supercord and ESAB's OK 46.00 are good choices, and are equally at home welding on the flat or vertically upwards.

Vertical down welds need more current than the same weld on the flat, as the extra force in the arc helps keep the molten pool clear of slag. Start with a setting close to the top of the rod's recommended range. Center the rod over the joint line (3.48) and keep it just below horizontal. Travel fast enough to stay ahead of the descending slag curtain. Don't let the arc get too long or weld metal won't find its way to the pool.

Welding **vertically up** produces deep penetration and operator fear in equal measures. But with the plant set right and the technique sorted, worry melts away.

Electrode choice isn't critical as it is with vertical down. Most general-purpose mild steel rods will work happily upwards, as gravity carries off the slag.

Heat control is everything. Keep the current down, dialing in 10%-15% less than on the flat. But don't go too low or the rod will stick and slag traps form in cool areas. It's a balancing act: experiment at the lower end of the rod's recommended range, settling on the lowest current compatible with trap-free, easy-running work.

While welding you'll see the arc digging a crater and molten metal draining away under gravity. This is where vertical up's fine penetration comes from. But if the area overheats, the end result is a series of dangling blobs like melted candlewax (3.52).

The first pass in a multi-run joint is made without weaving unless you need to control burn-though. Subsequent runs rely on a weave to support the molten metal (3.49). Each lateral weave creates a horizontal ledge, which is used as a

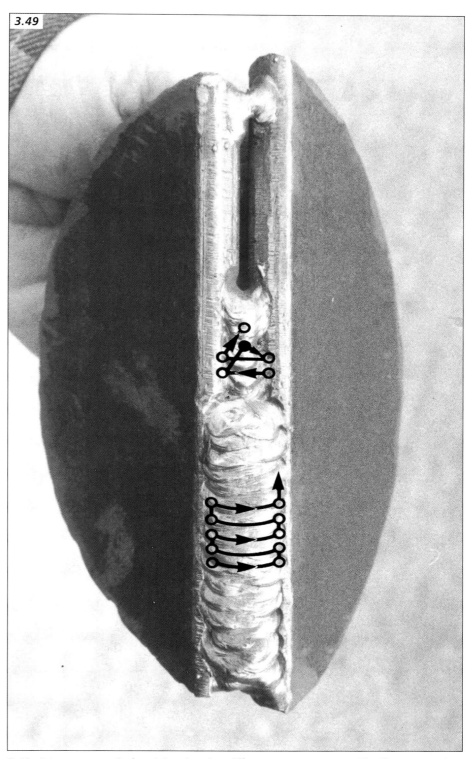

3.49

3.49. A two-pass vertical-up joint showing different weave patterns. The first run used a triangular weave, pushing the tip into the root gap until a change in arc noise indicated full penetration. The "keyhole" shape in the penetration gap confirms this (arrow). Pausing at the borders of the weave (circles) filled in the edges. The second pass features an easier H-pattern weave, again with edge pauses. But the second weave wasn't wide enough to blend with the plate's outer edges, and there is some undercut near the start caused by too-brief pauses.

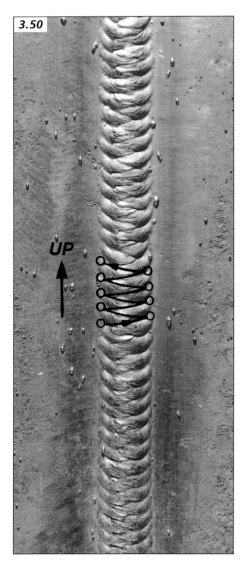

3.50. A simple crescent weave features in this quality single-pass run. Neatness and strength come from very deliberate rod tip movement, coupled with a pause at each edge to let the crater fill (circles).

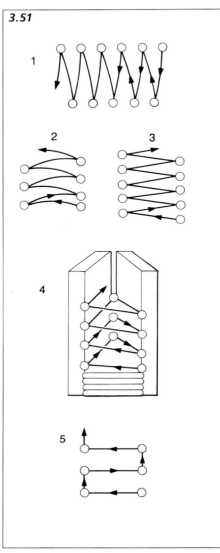

3.51. Assorted weave patterns. Use number 1 for flat (downhand) work, 2-5 for vertical up. Choose whichever best suits the job in hand.

3.52. Too much heat turns a vertical up weld into a mess. Gravity pulls molten metal down too far before it solidifies (arrow).

step to hold fresh weld. It's like making each rung of a ladder as you climb.

Follow the arrows in pictures 3.49, 3.50 and 3.51 to see where the rod tip goes during different weaves. A good upward weld has a wrinkled, prune-like finish; it's the result of all those individual ladder rungs. One or more downward passes using a crescent weave can tidy things up if needs be, but re-set the current and use the right rod (page 45).

For best weave control imagine you're a slow-motion robot, constantly tracing the same pattern on the joint at the same speed. Keep the rod horizontal or in thick sections, just above horizontal. Center it over the joint to balance fusion between the plates and hold arc length as short as current allows to direct metal where you want it.

If the set has more than one output voltage, use the higher line to help arc establishment and smoothness. Relax, taking the electrode cable's weight on

surrounding metalwork and propping yourself comfortably. When you're welding, try keeping a light grip on the electrode holder. Enjoy yourself! Repeat after me: vertical up is fun, vertical up is fun. . . .

Overhead welds are another tightrope act. It's flat welding turned upside down, only this time gravity is working against you. Metal will transfer across the arc thanks to magnetic forces, but the molten pool wants to drip. So three rules apply. Keep the arc as short as you can to lessen the chance of losing metal during transfer. Keep the current relatively low so weld metal can freeze before it sags; and travel relatively quickly to limit heat build-up. All this has to be balanced against achieving good fusion and penetration—hence the tightrope act. A visual clue comes from bead shape: if the center hangs down, you're a little too hot or too slow.

MMA Helpline

Everybody (but everybody) makes mistakes. There's no shame in it. But if you want to improve it helps to be able to spot faults, assess their seriousness and know where you went wrong. Pictures A–H deal with common problems and ways round them. (All pictures courtesy of The Welding Institute.)

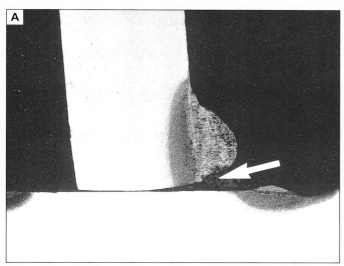

A. Slag traps (arrow) are a pain. Avoid them by keeping the four variables right — use enough current and the appropriate rod angle, hold arc length fairly short and don't travel too fast. The one above came from using too little heat or a long arc. Rod metal took the shortest path, leaving a hole for slag to fill. Dealing with traps is covered on page 44.

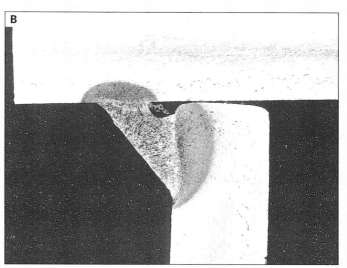

B. This weld seemed fine on the surface, but a peek inside shows unequal length legs and a slag trap instead of root fusion. Low current, a too-long arc and wrong rod angle were the cause.

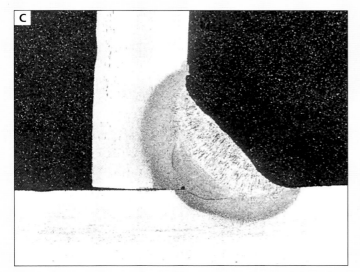

C. This T-fillet, welded with the right current, rod angle and arc length, shows equal leg lengths and good fusion in both plates. A second, weaved run was added — spot its coarse, grainy look. The first run has been normalized by the heat of the second, leaving its grains smaller and improving strength.

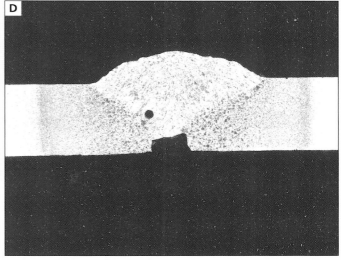

D. Two faults in one here: incomplete penetration despite the gap between plates, plus a circular slag inclusion at eight o'clock. A smaller diameter rod, run at a reasonable current, would have helped penetration and brought enough heat to prevent trap formation. Just winding up the current on a big rod wouldn't do, as its girth would still stop it reaching the bottom of the joint.

MMA Helpline *continued*

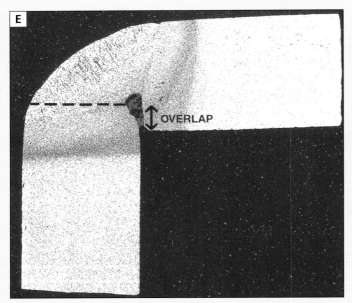

E. Poor preparation sank all hope for this outside corner joint. The plate edges were overlapped rather than kept parallel (dotted line), which kept arc heat from the joint root and allowed slag to flow in instead. Aligning the edges so there's no overlap and maybe leaving a small penetration gap is the answer.

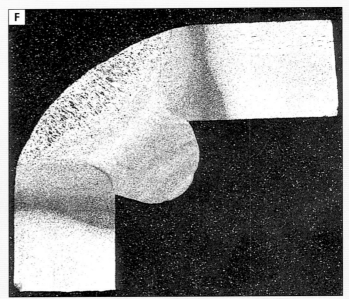

F. Another outside corner, only this time with too much penetration. Weld metal from the first of two runs has oozed right through the joint, either because the gap was too big, the current too high or travel too slow. The sudden change of section at the inner corners will concentrate stress, giving a site for breakage.

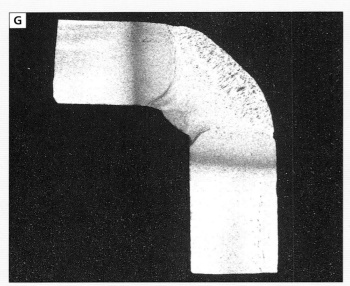

G. That's better. This T-fillet's two runs penetrate and fuse to full depth. The small hole at the root isn't a problem: it comes from the vertical plate's original rounded profile. If you're not convinced, lay a ruler along the right-hand edge.

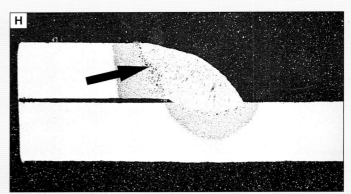

H. Nothing much wrong with this lap joint. But if it's loaded heavily, failure is likely in the heat-affected zone where grains are coarse and relatively weak (arrow). The torn surface would show a characteristic crystalline structure. There's not much that can be done, short of post-welding heat treatment to relive stress and normalize the grain. It's a common cause of unexpected breakage in welds that otherwise look good.

Section 4

MIG/MIG Welding

Now for something newer. MIG/MIG equipment was developed for industrial production, where its speedy work offset the originally fierce cost. These days, very capable sets are offered by a whole range of makers, so farms can easily exploit the process's benefits in fabrication and repair.

Here we'll deal with the sort of unit most likely to be bought new by an average farm workshop . . . whatever that may be. Typically it'll be a single phase set, running welding wire from 0.023"-.052" (0.6mm-1.2mm) diameter. A higher-capacity plant is likely to need three-phase power (which limits mobility) and costs more. Although such units offer extra control sophistication, their basic operating principles are the same.

So what is MIG/MIG, and why the peculiar name? It's an arc welding process in which the stick electrode is replaced by continuously fed wire, and where a shield gas takes over from an electrode's coating.

MIG is shorthand for Metal Inert Gas. But the only truly inert (i.e., non-chemically reactive) shield gases are argon and helium, which are reserved exclusively for welding aluminum or other non-ferrous metals and their alloys. So strictly speaking, MIG welding refers just to that.

Here, though, we're talking about welding steels, using either carbon dioxide (CO_2) or a CO_2/argon mix as the shield gas. As carbon dioxide is not inert, the process can't rightfully be called MIG. Instead, Metal Active Gas (MIG) welding is the proper term. So which to use? Although MIG/MIG is the proper moniker, it's a bit of a mouthful. So we'll settle for the common name — MIG — on the understanding that it's bending the truth.

Convert or Beginner?

MIG welding has a reputation for being easy, and it is if approached the right way. Yet it's equally easy to make big mistakes.

If you're coming to MIG from a stick welding (MMA) background, be prepared for change. Applying MMA rules to MIG just doesn't work, as the new technique stands most things on its head. For instance:

• MIG is always a DC (direct current) process.

• The operator is not in manual charge of arc length. This is changed by altering voltage at the set.

• Current is adjusted indirectly by altering wire feed speed.

• Travel is substantially faster and the gun is usually pushed, not pulled.

• There's effectively no help from flux.

The only really constant thing between MMA and MIG welding is the everlasting importance of penetration and fusion. That never changes, no matter whether the weld is made by MMA, MIG, gas flame or mind over matter.

MIG's Benefits

MIG welding offers a fair selection of advantages over MMA.

• MIG's filler wires carry similar current to MMA electrodes but are much thinner — for example 130A down a ½2" (1mm) wire, rather than though a ⅛" (3.25mm) electrode. So there's great potential for excellent fusion and penetration so long as technique is good.

• MIG is faster, in both speed of travel and post-welding cleanup.

• MIG operation requires less manual skill, as arc length is controlled by the set rather than by the operator. Against

this must be stacked the need for more careful setting of welding conditions, that is, the fine-tuning of voltage and current.

• The joint area is easier to see as there's no slag coat, so potential problems can be spotted sooner while you're welding.

• There's no slag to control.

• Given a reasonable volume of work, MIG is potentially cheaper — even considering the cost of cylinder rental and shield gas.

The last point has to be looked at carefully in a farm context. MIG plant is more expensive than equivalent MMA gear, thanks to its wire-feed mechanism and AC-DC rectifier. Extra complexity means more potential for going wrong. Thus buying a MIG set for occasional use makes little sense. But once you've got one, it'll probably take over the bulk of the work. Also, for some jobs MMA's wide electrode choice, (relative) contamination tolerance and immunity to nature's breezes make it the better choice. Small MMA sets (particularly inverter-based ones) are lighter and far more portable than MIG, too. Thus relying solely on MIG for the whole spectrum of farm work isn't a good idea, but MIG with MMA backup definitely is.

What's Involved — More Detail

MIG uses bulk electrode wire on a spool. The user holds a gun-like torch, whose trigger starts and stops wire feed, welding current and shield gas flow. Real bargain-basement sets leave the welding wire live all the time, so there's always the risk of accidental arcing. Don't bother with these.

Wire is fed continuously to the weld pool by a motor and roller arrangement, with feed speed varied by a control on the set (4.1). The wire passes through a copper

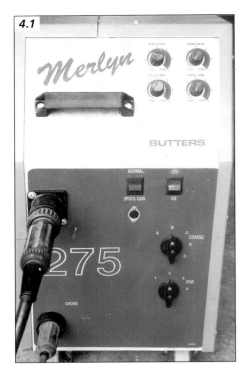

4.2. Three ways to control shield gas flow. Budget MIG/MAG welders often come with a simple flowmeter (left). The single-stage regulator (center) shows cylinder pressure on its dial and gas flow on its graduated barrel, while a two-stage regulator (right) adds precise control over delivery pressure. For normal work, a single-stage regulator is a good compromise between cost and control. Note: Any regulator must match the gas it is used with. Oxygen equipment is shown here simply as an example of regulator design.

4.1. This upmarket single-phase MIG set has click switches to change voltage in coarse and fine steps, plus a rotary knob for wire speed. Other rotaries look after burn-back control (how long the arc stays on after the trigger is released, and thus how far wire burns back) and arc on/off times for spot and stitch welding. The spool gun socket (center) is used with a remote wire feed unit.

contact tip as it leaves the gun, picking up electrical power en route. A copper coating on the wire improves electrical contact and delays surface rusting.

Compressed gas shields the weld pool. A regulator or flow meter on a bulk supply cylinder (4.2) varies gas delivery pressure, passing it up to the gun through a flexible tube. Flow is started and stopped by an electrical solenoid valve activated by the gun's trigger.

The gun — or torch — ends in a nozzle. Usually height-adjustable, this directs the gas over the weld area to hold off atmospheric contaminants (4.3).

How about the filler wire? Apart from steel, it includes small quantities of deoxidants (silicon, manganese and perhaps aluminum, titanium and zirconium) to help counter the effect

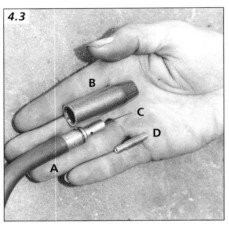

4.3. Shield gas flows from the gun (A) before passing through the nozzle (B), and then to the weld area. Filler wire (C) is fed to the weld pool mechanically via the torch liner, picking up electrical energy from a copper contact tip (D) as it leaves the gun.

of shield gas-derived oxidants on the weld pool. But these can't match a MMA rod's flux coat in cleaning effectiveness, so MIG is much more contamination-sensitive than stick welding.

The set's front panel carries more controls than a stick welder. Along with the on/off switch you'll find either push buttons or click switches to change voltage, and a rotary knob to vary wire speed. More complex sets offer functions

like spot weld timing, burn-back control and variable inductance. Page 52 has more on these.

Voltage and Wire Speed — An Intimate Relationship

Before going into the nuts and bolts of setting up a MIG set, we'll see what happens when you change voltage and wire speed. Understanding how these two interact is important, and there's more detail on pages 51-52. But if you don't fancy that and if just contemplating electricity gives you the willies, here are the basics.

CHANGING VOLTAGE

With MIG, arc length and voltage are directly related. The more voltage you dial in at the control panel, the longer the arc gets. Think of it as a hosepipe being lifted over a vegetable patch: the more you lift, the longer the water jet gets and the wider it fans out. If you keep on increasing voltage without stepping up wire speed, the arc eventually ends right back at the contact tip, fusing the filler wire in it — have a go and see. To recap: more voltage produces a longer arc, which heats a wider area. Dropping voltage shortens the arc and heats a narrower strip.

CHANGING CURRENT

There's no knob on a MIG set to vary welding current. Instead you do it indirectly by altering wire feed speed.

But how? The power unit is designed to hold voltage (arc length) to the operator's pre-set value, but is free to vary current within limits. At any instant, it delivers just enough current to melt filler wire at the rate needed to hold arc length steady. If you increase wire speed, the set delivers more current to keep the wire burning back at the same rate and thus holds arc length the same. If you slow wire speed down, the set pegs arc length by delivering less current so the wire doesn't burn back as fast. Thus changing wire speed varies the (average) current.

But only within limits. If you keep lowering wire speed, there comes a point where the set can't compensate any more. From here on, wire will be burnt back faster than it's fed to the weld pool and you'll notice arc noise and appearance change. On the other hand, if you wind up speed to the point where the set can't deliver enough extra current to burn off the incoming wire, that wire starts to rush into the weld pool more or less unmelted. You'll feel this through the torch as juddering or "stubbing" as it hits the underlying plate. To sum up: *voltage has the biggest effect on arc length; wire speed affects it less. Changing wire speed alters average current, though the available variation isn't huge.*

Practical points to emerge from all this are:

• In MIG, the operator preselects voltage and thus arc length.

• Higher volts give longer and wider arcs. Once wire speed and diameter are chosen to exploit the higher energy available, thicker sections can be welded.

• Lower voltages give shorter arcs and narrower bands of heating, so are used with thinner wires on thinner work.

• At any given voltage setting the operator can vary average current via wire speed. Faster feeds mean more current, slower feeds mean less current.

Thinking Caps On

To really get the most from MIG it's necessary to twig how the process works. Figure A shows what's going on in the weld zone. A key point to start: a long electric arc needs more voltage to sustain it than a short one. Why? There's more electrical resistance in a long arc, which takes more volts (electrical pressure) to overcome. As arc length grows, voltage goes up and current drops.

MMA and MIG power supplies are quite different animals, with the way they work making it possible to control the two forms of welding. We'll start by looking at what goes on inside a stick welding (MMA) plant. The user sets welding current to give the required heat input, then fine-tunes this manually by altering arc length. The set's voltage output is fixed at one or more levels depending on the equipment, with 50V and 80V lines common.

For consistent fusion and penetration, welding current needs to remain as

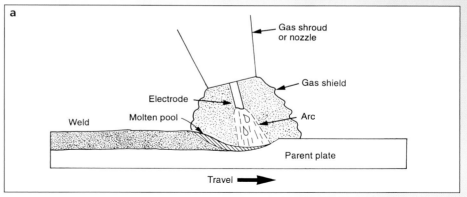

A. MIG in action. Shield gas keeps air from the molten pool, mechanically fed wire is the electrode. Filler metal can get to the pool is several ways, giving practical implications — page 55 explains.

constant as possible during work. But current varies with arc length, and the MMA operator's more-or-less unsteady hand means that arc length is always shifting a little. As it lengthens, arc voltage goes up. As it shortens, voltage drops. So welding current wants to change too.

The stick set's internals try to minimize current change. They are designed so that a large change in arc length (and

thus voltage) produces only a small change in current. This minimizes the effect of operator unsteadiness on welding heat. The power unit behaves like an elastic band between voltage and current: hold one end steady (current) while you pull the other by changing arc length (voltage). Figure b puts this in graph form.

MMA electrodes burn back fairly slowly, giving the operator time to spot and counter changes in arc length. But when he's half-asleep, the power supply helps by holding current as steady as it can — which minimizes the effect on weld quality and rod burn-back until the operator wakes up again. Despite this, MMA welding still largely depends on the user's ability to feed rod steadily to the joint.

(As an aside, you might fall to wondering — in a bored moment — why an AC welding set can work at all. After all, AC voltage and current cycle repeatedly through zero, which ought to stop the arc. The day is saved by reactance in the power unit. This electrical phenomenon introduces a designed-for lag between AC voltage and current. As a result both don't fall to zero together, and the arc keeps running. MIG sets are always DC so don't have the same potential problem).

MIG Power

By contrast MIG's thinner filler wire burns back far faster than a rod, to the point that even Superman would be hard pushed to keep arc length constant. So it needs an automatic control system.

This small miracle is achieved by reversing the "elastic band" effect inside the power unit. Voltage is pegged while current can be stretched to and fro. If the operator lifts the torch from the work, arc length tries to grow and voltage wants to climb. But the power unit is designed so that as soon as voltage shifts, current changes even more. Compare Figure c with Figure b to see what's happening.

What does this quick-shifting current achieve? Think again about the operator lifting the torch. Arc voltage tries to go up, but current drops faster. The filler wire immediately burns back more slowly — and as it's being fed

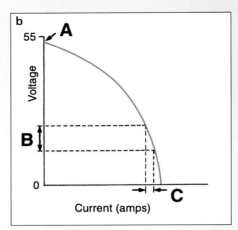

B. Power supply characteristic for MMA. On striking the arc, voltage falls from the open circuit value (A) to the running voltage band (B). Alterations in arc length produce voltage variations in this band. See how the curve's shape produces only corresponding small changes in welding current (C)? The set actively helps to maintain current — and hence weld quality — as the operator's arm wobbles.

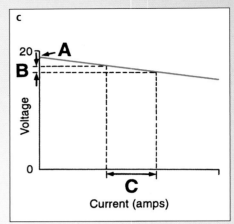

C. The MIG power supply holds voltage and thus arc length (B) relatively constant for a large change in current (C). As the operator wobbles, arc length tries to shift. But the rapid change in current burns off more or less wire, returning length to where it was. This makes MIG intrinsically easier to use than MMA.

mechanically at a constant rate toward the weld pool, the arc gets shorter. This shorter arc needs less voltage to run, so the power unit raises current a little to burn back the wire faster. The result is an arc that stays on average constant in length, self-adjusting quicker than you can notice.

Automatic arc adjustment also brings a bonus. A longer or shorter arc would fan out more or less on the work — remember water spraying from a hose. Left unchecked this would increase or decrease weld pool width, but MIG's self-adjustment tends to keep pool and bead width steady.

Some MIG sets also offer **slope control**. By changing the slope of the power supply's volt/amps curve (c), this gives a way to fine-tune spatter. At any given working voltage, a shallow slope lets short-circuit current rise more than a steep one does. High short-circuit current practically explodes the molten drop from the wire tip,

while lower current lets it part more slowly. So dialing in more slope limits short-circuit current, and thus reduces spatter. Lower cost sets end to have slope fixed.

You might also find **variable inductance** on more capable sets, usually achieved by plugging the return lead into alternative outlet(s) on the front panel. This controls the rate of current rise during dip transfer's short circuits, with more inductance giving a slower rise and gentler pinching of droplets from the welding wire. One result is less spatter; others are a longer "arc on" time (and thus a more fluid weld pool) and a flatter, smoother bead. If you have it, use variable inductance to control spatter and improve finish: but bear in mind that too much inductance for a given voltage produces poor starting. Experiment in both dip and spray modes to see its effects.

WHICH WIRE?

Smaller single-phase sets handle at least 0.023"-.030" (0.6mm and 0.8mm) diameter wires, while bigger-capacity sets take ⅟₁₆" (1.0mm, 1.2mm and maybe 1.6mm) sizes as well. Keep stocks of all your set can use, buying big-name brands for reliability.

Wires come in different grades. For fabrication or repair in mild and low-alloy steels, double-deoxidized wire to grade ER70S-6 of the appropriate AWS is fine. Note the low-alloy coverage: using standard wire, MIG can deal with slightly special steels like weld-on power harrow tips. Unless you specify otherwise, ER70S-6 is what your supplier will probably provide. If you don't mind paying a little extra, precision-wound spools give a smoother feed than random-wound versions.

Triple-deoxidized wire to AWS ER70S-7 handles contamination better but is significantly more expensive. Other wire grades cover high-alloy and stainless steels but these don't feature much in farm work.

Steel filler wire is generally copper-coated to resist corrosion. Prolonged damp affects performance. MIG wire has a big surface area in relation to its diameter, so even light surface rust interferes with smooth feeding and produces lower-strength welds. If the set won't be used for a while in winter it's best to take off the wire spool and keep it somewhere dry — spare spools should always be stored this way.

Clean filler wire always feeds and welds best. A simple fit-up with wire wool, cloth and a couple of clips will scrub the filler, helping feed rollers get a grip and minimizing contaminant build-up in the guide tube (4.4a, b and c).

WHICH GAS, AND HOW MUCH?

For mild steel, use either CO_2 or a CO_2/argon mix.

CO_2 gives narrower, deeper penetration and a less fluid molten pool that's easier to control on positional work.

4.4. In winter damp the top coils of MIG wire pick up a thin, patchy layer of rust. Steel wool and a piece of rag help produce a smooth feed and consistent work (a). Don't clamp the wire too tight, and change the pad material occasionally. If wire feed is erratic look at spool friction drag (b) and feed roller adjustment (c). Wire drives come with alternative rollers to match different filler diameters.

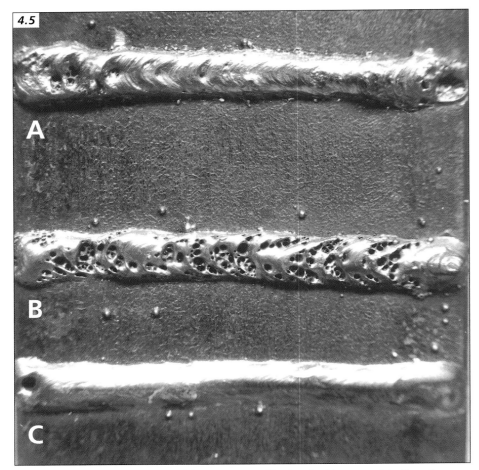

4.5. Shield gas effects. No gas (A) leaves mousse-like bead with some pinholing. Too little gas flow (B) gives heavy pinholing: weld zone contamination gives the same effect. Adequate gas flow (C) shields the weld completely.

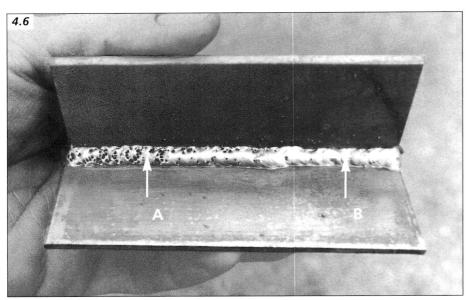

4.6. Contamination produces a weld that looks like a Hershey's Air Delight chocolate bar and has about the same strength (A). This CO_2-shielded weld improved as the plate got cleaner (B).

It's cheaper; a cylinder lasts longer and the extra penetration can be a boon on thick section steels. But CO_2 produces a rougher weld finish with more spatter. The arc is harsher and spray transfer can't be achieved. Globular transfer is the norm so spatter can't be tuned out by the operator. Ideally, triple-deoxidized wire should be used with CO_2 shield gas to counter its oxidizing tendency and thus maximize weld strength.

Argon/CO_2 mixes give a hotter, more fluid molten pool and support true spray transfer. Penetration is broader but shallower, making gas mixes better for thin-section material. Spatter is reduced, the weld surface is smoother, the arc more stable and plant setting is less critical. Gas mixtures with 5% or 20% CO_2 are on offer in most markets, though not necessarily in all cylinder sizes. A 20% mix gives a slightly cooler pool better suited to positional work and helps penetration in thick sections, while the 5% variety is best on thinner stuff. For general farm work, a 20% gas mix is good.

Gas supply rates are often quoted, with 25-30 cubic feet per hour (10 liters/min) or 3-5 psi typical. In practice it's enough to use your ear alongside the nozzle. Set wire speed set to zero (so you don't get earful of wire), pull the torch trigger and increase flow at the regulator until a steady, soft hiss is heard. Then, once work has started, decrease flow as much as possible. Too little shield gas gives the weld a rough, mousse-like look: try short welds with different gas flows to see the effect (4.5). Note that contamination in the weld zone also leaves a poor finish, usually riddled with holes (4.6).

When welding in a draught or at high voltages, increase gas flow. Using a sheet of tin to keep the gale off the work area will help, though MIG can't cope with much airflow; the gas shield is blown about or vanishes. But in normal work, using "a little more gas to be sure" is just a waste. The shield gas has absolutely no effect on the arc's ability to handle rust and other contaminants, so there's no point in using more if the arc isn't behaving. In such cases stop, find the problem and put it right. Using too much shield gas runs the risk of actually lowering weld protection (especially with CO_2) as the stream becomes turbulent, drawing in the very air that you're trying to exclude.

MIG Setup 1: Dip or Spray?

MIG sets are a bit like Janus, the two-faced Roman god. Looking one way is an inviting "easy-to-use-me" smile. Pointing the other is the more mysterious side, labeled "setting up." But twiddling the knobs to get the best from MIG isn't too hard once you've got a grip on the principles.

Unlike with MMA, you get a choice of how to deposit the filler metal. The mode used — either dip transfer or spray transfer — has big implications for pool control, fusion and penetration. Whether there actually is a choice depends on the set's maximum output, the filler wire diameter and the shield gas.

LUCKY DIP . . .

Where work involves either thin sections, positional (i.e., not flat) work or gap filling, **dip transfer** is the mode to use. In this there's not enough voltage for the arc to stay alight constantly, so it dips in and out of a weld pool that's relatively small and fast-freezing. Heat input is relatively low (so distortion can be too) and filler metal is only transferred while the wire is in the pool, never in free flight across the arc.

Illustration 4.7 shows dip transfer. Filler wire comes out of the torch's contact tip just faster than the arc can burn it off, so travels down and touches the work. The resulting electrical short heats the wire until a length melts into the weld pool, establishing a temporary arc. The process repeats 20-200 times/sec, generating MIG's characteristic crackle.

. . . UNLUCKY GLOBULAR . . .

As voltage and current (wire speed) are increased, the way filler wire metal is transferred to the weld pool changes. More voltage allows a self-sustaining

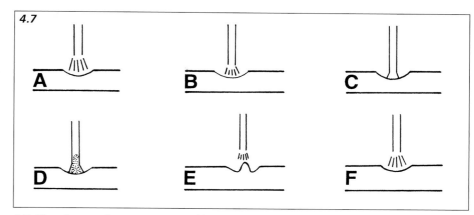

4.7. The dip transfer sequence runs like this. An arc exists between wire and weld pool (A). But as wire is feeding faster than the arc burns it back, the tip gets closer to the pool (B). When wire and pool meet the arc goes out (C), producing a short circuit. Current flow heats the wire (D), which melts to leave a globule in the pool (E). The cycle is then back at square one (F). MIG's characteristic crackle comes from the on/off arc.

arc, and dip transfer shifts to globular transfer. In this, large blobs of molten wire move across the arc in free flight (4.8). Droplets are the same width or wider than wire diameter and break off randomly, changing the smooth rip of arc noise into a rougher popping.

Although heat release is greater than in dip transfer (as the arc is "on" continuously), globular transfer is not very good news. There's not enough magnetic energy to direct those big droplets to the weld, so either they miss it altogether or whack home like a hippo jumping into a paddling pool. Either way, metal is lost from the joint and spatter goes everywhere.

There is no off/on transition between dip and globular transfer, just a blending from one to the other. Where in the wire speed range this happens depends on the set and shield gas, but it'll be at higher feed speeds.

If you're using an argon/CO_2 mix, side-stepping globular transfer is a matter of spotting what's happening and tuning it out — see pages 58-60. But if you're using pure CO_2 you can't avoid globular transfer. Penetration will be good but the higher the current goes, the worse the spatter will get and the rougher the weld will look.

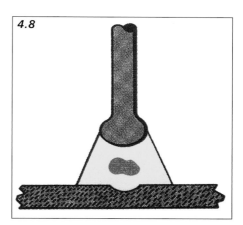

4.8. Globular transfer. When voltage and wire feed speed (current) are both high enough, a self-sustaining arc appears. But energy in it is not high enough to produce true spray transfer or channel molten filler metal straight to the weld pool. The result? Blobs of filler fall through the arc into the weld pool, or are pushed sideways and out. Spatter is high, weld finish is poor.

. . . AND SUPER SPRAY

Spray transfer is the third way to deliver filler metal to the weld pool, though it's available only at higher arc energies and when using a mixed shield gas. Spray can't be achieved with CO_2.

With smaller wires, the set must deliver at least 160A and the shield gas must have 80% argon at minimum. Thicker wires need higher energies still.

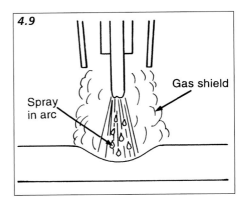

4.9. Spray transfer. Higher arc energies pinch off filler wire in much smaller droplets, which are taken down through the arc by magnetic field. Spatter is low, the arc long and wide, the weld pool hot and liquid. Penetration and fusion are potentially high.

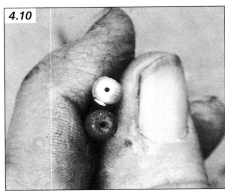

4.10. Compare a worn contact tip (bottom) with a new one. Change as soon as wire becomes a floppy fit, or after spatter damage. The wire must be a free-sliding fit, never tight.

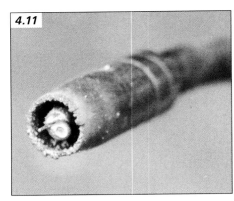

4.11. Dip transfer at high voltages leaves a crusting of spatter in the nozzle, particularly when working vertically or overhead. This eventually cuts gas flow and can bridge to the contact tip, turning the whole nozzle "live." Check and clean frequently.

4.12. Cheap and very useful, MIG pliers put a nozzle reamer, wire cutters and tip spanner in one tool.

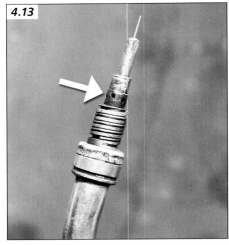

4.13. Torch gas holes (arrow) can block with spatter, limiting flow. Using too much release spray can make things worse.

For a given wire diameter, combining high voltage and high wire feed speed will produce globular transfer. But by winding back the wire feed speed a little, the wire can be burnt off in a more orderly fashion. Much smaller droplets will pinched off the wire's end, to fly across the arc in a fine spray (4.9).

The long, high-voltage arc delivers heat in a wide fan. Fusion and penetration go up compared to dip transfer, suiting spray to heavier steel sections. The downside is that the hot, fluid weld pool limits work to the flat (or gently sloped) position and is too fierce for thinner materials. And keep an eye on the torch when spray is used, as the extra heat released can cook it.

Noise changes as globular transfer turns into spray. Its uneven stutter fades to a smooth, low hum. Light and heat output both go up noticeably. You can't miss it.

MIG Setup 2: Equipment Checks

There's a lot to look after with a MIG set, so here's a pre-welding list.

Contact tips are consumables. They're cheap and don't last forever (4.10). The wire picks up welding current here and good, consistent contact means a smooth arc. When wire flops about in a used item, fit another. If a tip looks

so-so, try slipping a length of the next-fattest wire in it: if this goes through, throw the tip away. Spatter can part-block the exit hole which slows the wire, and there must be good, easy-sliding contact between the two for an even feed. Unless you're desperate it's not worth trying to clear a blockage with gas nozzle reamers — the soft copper is rasped away before harder spatter, leaving an oval, oversize hole.

If the **nozzle** crusts with spatter, less shield gas will flow (4.11). Really serious crud will bridge between nozzle and tip, turning the nozzle live and risking an

arc between it and the surroundings. Clear spatter with a blade or special tool—you can buy pliers which combine wire cutters, tip pliers and reamers in one unit (4.12). Spatter-release spray helps combat build-up, but don't let it block the gas exit holes below the contact tip (4.13). Looked after, a nozzle lasts a long time. During work check its condition frequently.

Adjust **nozzle height** to suit the welding process (4.14). For dip transfer, the contact tip can be level with or just peeking out beyond the nozzle. This minimizes electrical resistance caused

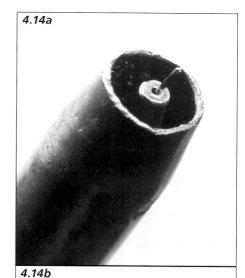

4.14a

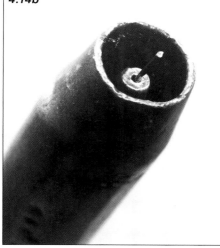

4.14b

4.14. For dip transfer, set the nozzle so it's level with or just below the contact tip (a). This gives the best view of proceedings. For hotter spray transfer, set the nozzle so it masks the tip by about ¼" (6mm) (b).

by long stick-out (page 62), while maximizing welding current and the operator's view of the proceedings. When working in a tight corner try exposing the tip even more, but spatter build-up will be fast — as it will when working vertically or overhead. Be careful in dip transfer. The more the tip is recessed, the greater is the effective stick-out and the more you'll be tempted to make up for lost current by increasing wire speed. Bead shape and penetration then suffer.

For spray transfer draw the nozzle forward so it shrouds the tip by about

¼" (6mm) (4.14b). The tip is then held back from arc heat and the nozzle can direct shield gas for better weld pool protection.

Spare a thought for the **torch liner**, the replaceable internal spiral which guides welding wire. An occasional blast-out with dry compressed air clears its tubes, giving a smooth wire feed and steady arc. Use no oil! Even a good liner won't last forever, so don't overlook it if wire feed turns nasty.

The spool spindle has an adjustable **friction control** (4.4b). Set this so the spool won't overrun when the feed motor stops. Too much friction loads the motor and can mean a jerky feed as rollers slip and slide. Tighten the damper before moving the set around in a vehicle, or the bumping will produce a bird nest of wire.

Too much **feed roller pressure** can damage the wire, yet too little gives an uneven feed. The latter shows up as an unsteady arc and (perhaps) in the tendency for wire to burn back and fuse with the contact tip even though voltage is right. Back off pressure until the rollers just start to slip as you brake wire between finger and thumb; then increase it until the slipping stops. Keep the rollers clean and free from oil — it's surprising how quickly crud builds up on them.

MIG Setup 3: Finding Good Welding Conditions

You've decided on the filler wire, the shield gas and picked a metal transfer mode to suit the job: dip for thinner stuff and positional work, spray for heavy sections if welding position allows. Now to set the set.

The idea is to balance voltage and wire speed to produce a particular transfer mode, then tweak these settings for the job in hand. Several things are important.

• There is no spot-on right setting. Instead, there's a narrow band of **good welding conditions** to find and move around in.

• At the borders of good welding conditions are transition zones into worse conditions. Steer clear of these.

• The calibrations on the set's controls seldom tie in with any specific voltage or wire speed. They're just markings, useful as waypoints during setting and for subsequent recall.

• Although there are tables laying down the minimum and maximum currents usable with each wire size, these are not much good unless the set has read-outs for volts and amps. Older and/or cheaper sets don't run to such things, so unless you rig up your own meters you must fly by the seat of your pants. Table 4.1 has ballpark values for those who can use them.

• How do you know when to change wire size? There is no hard rule. Just economics, plant capability and common sense. As joint size goes up, more wire is needed to fill it. Shifting to a fatter wire will deliver the extra metal while keeping feed speed down, forward travel rate up and holding to the required transfer type. Thicker wire (at the same voltage/current) also means less penetration and heavier weld deposit. At the other end of the scale, the lower usable limit for a given wire comes when the voltage (arc length) needed to burn it off blows holes in the work.

CONDITIONS FOR DIP TRANSFER

It's possible to come at good welding conditions from two directions. The precise way is to set wire speed to a given value and then use tables and experiment to find the right voltage to burn it off, which is ideal where all the variables are known.

But in the suck-it-and-see farm workshop it's easier to pick a voltage that delivers the heat you want, then find the wire speed range to suit it.

Until you know your set inside out, tune up first on a bit of clean scrap steel of the same thickness as the job. Dip transfer is a low arc-energy mode, so pick a voltage toward the lower end of the scale and set wire speed to about

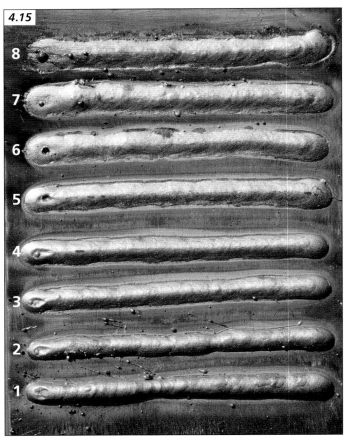

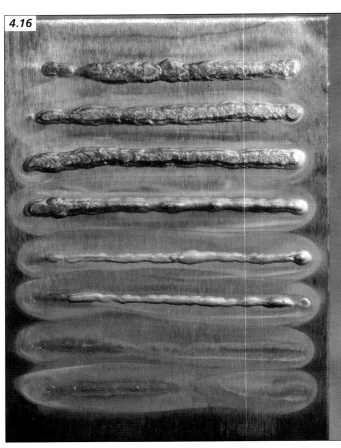

4.15. The voltage effect. In runs 1 and 2 voltage is too low; the weld bead is narrow and fusion poor. Increasing voltage through runs 3-6 produces good welding conditions: the bead is widening and flattening, fusion is fine and spatter disappears. Run 3 shows cool dip transfer, run 6 hot dip transfer. By runs 7 and 8 there was not enough wire to satisfy the lengthening arc, which burnt progressively back towards the contact tip. The bead is wide and very shallow, undercut appears and spatter comes back as globular transfer creeps in.

4.16. The reverse of 4.15. Moving up from low voltage (bottom) to high shows increasing penetration.

one-third of maximum. Adjust voltage until you're happy with bead width and penetration, using picture 4.15 as a guide. Then fine-tune current with the wire speed control (4.17). Two important points: while it's okay to alter wire speed while the arc is running, don't change voltage as there's a good chance that you'll blow the set. And to minimize confusion, only alter one variable at a time.

Here's the process:

1. Adjust voltage.

Pictures 4.15 and 4.16 show a series of runs on 0.08" (2mm) plate, made with 0.030" (0.8mm) wire and 20% CO_2 shield gas. The only thing changed between each run was voltage; wire speed was left alone. Each run corresponds to one voltage step on the set, starting low and working up. See how bead shape and spatter level change?

In MIG, voltage controls arc length—the more volts you dial in, the longer and wider the arc grows. In runs 1 and 2, voltage is too low to burn off the incoming wire. The bead is thin, uneven, humped and shows little sign of fusing with the parent plate as heat is lacking. There's a lot of spatter from weld metal being physically splashed from the pool, and the operator could

feel "stubbing" through the torch as wire jabbed down into the plate.

With increasing voltage, runs 3-6 blossom into good welding conditions. Now there's enough energy to burn off the wire, so the bead is regular and fuses to the plate. With each step up in voltage, the arc lengthens and widens a little so the bead flattens. By runs 4 and 5 dip transfer is in full swing, the arc is sizzling nicely and spatter has died clean away.

By run 7 the party is nearly over. There's so much voltage that the arc is burning well back up the wire. The wider arc produces a wider melt zone. And could

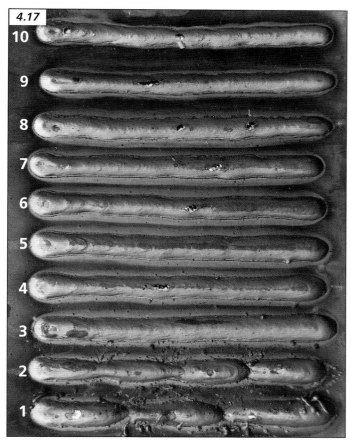

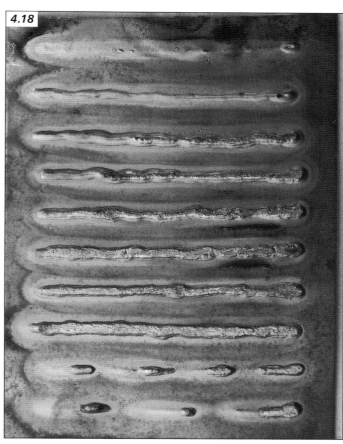

4.17. The effect of changing wire speed (current) while voltage stays constant. Speed starts off high (bottom) and is gradually slowed. In runs 1 and 2 speed is too high. Runs 3-9 are all good welding conditions, run 10 looks OK on the surface but has no penetration; see text for explanation. All runs are dip transfer with 0.030" (0.8mm) wire, argon/CO_2 shield gas on 0.08" (2mm) plate.

4.18. The flip side of 4.17 shows how a drop in wire speed reduces current — and with it, penetration.

you but hear it, your ears would tell you that the rate of sizzle is slowing right down.

In run 8 (maximum voltage) the arc is burning right back to the contact tip in a series of fuzzy-sounding "whooshes." It's so hot that the melt zone is wider than the filler could fill, leaving undercut at the edges. Globular transfer has started, so spatter is back. The whole area has a powdery brown coat.

This bad-to-good-to-bad sequence happens at any wire diameter as voltage is wound up. Picture 4.16 reinforces the point. Turning over the plate shows how penetration rises and widens with voltage. Using the above as a guide, take a bit of plate and run the experiment

yourself! Then pick a voltage that suits the work.

2. Adjust wire speed (current).

Once you've settled on a good voltage, bring wire speed into balance with it so penetration and bead shape are right. We've already seen in theory how current goes up with wire speed: now 4.17 and 4.18 will confirm it. These show a series of dip transfer runs in which wire speed starts fast, then is slowed a little each time. Voltage setting stays the same throughout.

When wire speed is too high for the voltage (runs 1 and 2), the arc can't burn off incoming wire fast enough. The torch jumps, the arc stutters and pops, spatter flies everywhere. The weld bead

(such as it is) is humped and pinched, coming and going between bouts of wire stubbing.

By run 3, wire speed has slowed enough to maintain a more or less steady arc. Spatter has almost gone and the weld bead starts to look respectable. Fast feed means high current, so fusion is good; check the corresponding penetration in 4.18.

Lowering feed speed progressively through runs 4-8 sees the bead gradually narrowing as current drops, but welding conditions are still good. Arc noise softens and the rate of ripping crackle slows with it.

By run 9, current is getting low. Dip transfer is still happening but dip rate is

4.19. Wire speed adjustment involves fine tuning. On the Butters set used to weld these examples, 4.15's full range of welding conditions was delivered in the narrow arc between H and L. Other sets are calibrated differently . . . which is all part of the joy of welding.

slow. The arc comes and goes, the bead is noticeably narrower and starting to hump as heat input drops.

In run 10 dip transfer has more or less stopped. The arc is burning back up towards the contact tip and wire falls as blobs into the pool, so the narrow bead's width varies as a result. From here, all it takes to get back to good welding conditions is a nudge up in feed speed.

Turning the plate over (4.18) shows how wire speed (current) affects penetration. After steady dip transfer is established (everything below runs 1 and 2) current goes down as feed speed drops, so penetration declines. By run 10 it's gone altogether.

Depending on the set, these changes will happen in a bigger or smaller arc of speed knob movement (4.19). Run the sequence for yourself— the examples used 0.08" (2mm) plate, .030" (0.8mm) wire and 20% CO_2 gas. With quality sets like the Murex Tradesmig and Butters Merlyn used to produce the samples shown here, small changes in feed speed deliver very definite results. With a cheapo set, adjustment effects can be less clear-cut.

Summary

The foregoing shows that once you've settled on a voltage that gives enough fusion for a given job, more wire speed delivers more current and penetration.

It's also clear that there's a definite band of good welding conditions (bounded on either side by settings that deliver duff results) and that it's possible to move around the good band by tweaking wire speed.

Pick voltage and wire speed to tailor a weld to the job, knowing what your own set can do. For instance, to produce a flatter bead with minimum surface build-up, use a higher voltage and lower wire speed to give hot dip transfer. For a narrow bead on thin sheet joints, limit arc length by lowering voltage and keeping feed speed down to minimize current — cool dip transfer. Or to reduce bead volume in a particular joint, increase voltage for a set wire speed or reduce wire speed for a given voltage.

CONDITIONS FOR SPRAY TRANSFER

Spray transfer involves higher energies and produces more light and heat than dip transfer. Allow for these by using an 11EW mask filter at minimum, and adjusting the torch nozzle so the tip sits about ¼" (6mm) up inside it.

If you're using 0.030" (0.8mm) wire and a 160A-180A set, start by dialing in maximum voltage. Thicker wires will need more volts, so smaller-capacity sets may not deliver spray with them. To see if spray is happening run a short bead on a piece of scrap, which needs to be at least ¼" (6mm) thick to take the heat.

At these high voltage levels the physics of the arc/wire system won't allow dip transfer. So if the arc crackles hard and there's a lot of spatter about, globular transfer is going on.

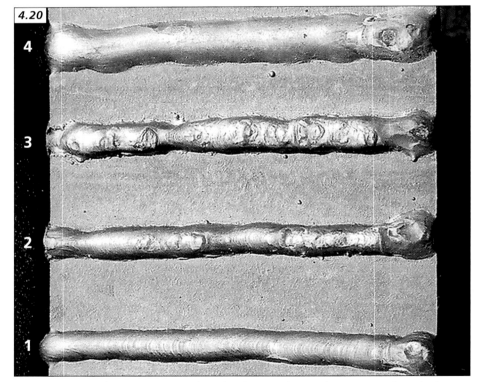

4.20. Three transfer modes, all produced with 0.04" (1.2mm) wire on ¼" (6mm) plate. Run 1 is low-voltage dip transfer; cool, orderly and spatter-free. For run 2, wire speed and voltage were both increased until globular transfer appeared. Although sounding something like dip, the arc's smooth crackle harshens and shifts to a more staccato popping as metal transfer becomes disordered. A lot more metal leaves the arc as spatter — check the bead area. Run 3 shows spray transfer, generated by slowing wire speed until the arc lengthens and fans out. Noise changes to a smooth, quiet hum and light output increases. The jump in heat produces a wide, smooth bead. Orderly metal transfer across the arc produces very little spatter.

To move into spray, lower the wire speed a little and try again. Keep lowering it until the arc suddenly widens and lengthens: at the same time light output will jump, arc noise shifts to a quiet humming and spatter dies away as metal transfer becomes more orderly. Now you've hit spray transfer (4.20).

The arc fans out to generate a wide, fluid and slow-freezing pool. If this is too wide for the job, drop wire speed just a touch. The arc will narrow and start to crackle faintly. This tighter spray is good for fillet joints, as it puts more metal into the area where the two plates meet.

Check penetration and fusion. Both will be much better than with dip transfer. If they're excessive for the job, try a lower voltage. But you can only go down so far and maintain true spray conditions. When voltage drops too low for the wire in use, transfer will move back to globular mode. Spatter will increase substantially and the arc will splutter.

CONCLUSION?

All this business of setting one thing and adjusting the other looks complicated on paper, and sometimes isn't that much better when you've got the torch in your hand. But persistence and experiment pay off. Settings are never quite the same on different jobs; get to know what you want and what you're seeing and hearing, then fiddle until the set delivers the goods.

Watch Out: Dodgy Combination About

Careful plant setting and good technique will produce a fine-looking weld, but you still need to think about what's going on. A joint can look a treat but have the strength of a stick of Brighton rock (a and b). (If Brighton rock is rare in your part of the world, substitute your favorite brittle confectionary.)

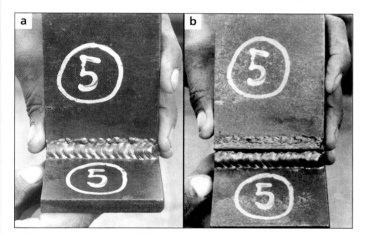

What's wrong with this fillet? To judge by the bead surface (a), not much. But when stressed, the joint failed suddenly under light load. See how little fusion there was in the vertical plate wall (b), and none in the root? Thin filler wire, delivered by dip transfer, couldn't put in enough heat for good fusion/penetration.

4.21. A subsoiler wing, welded with spray transfer using a 180A set and 0.030" (0.8mm) wire. Dip transfer wouldn't have produced the required penetration, given the foot's big heat sink effect.

Getting penetration takes heat. The thicker the section, the more heat it absorbs before melting. So for a strong job in thick sections, the arc must deliver enough energy to satisfy the steel's heat-swallowing tendency and still have enough left over for a good, deep melt in the weld zone. With MIG, voltage is arc width and wire speed is current. Only the bigger wire sizes can handle the voltage and current needed to guarantee penetration in thicker stuff.

So the message is simple. If you try welding heavy sections with small wires and dip transfer, low energy input can produce a joint that looks good but is inherently weak. Are there any options if you're stuck with a small set and limited fillers?

- Restrict 0.023" (0.6mm) filler wire to material under ⅛" (3mm) thick.

- With thicker-section steel use the highest possible voltage that still maintains dip transfer; don't edge into globular. Use reasonable wire speed to keep current (and thus penetration) up.

- Travel at a speed which keeps the arc at the front edge of the weld pool. This is a balancing act. Work too slowly and the arc fires more into the weld pool than the parent plate, so penetration is reduced. Work too fast and there's not enough time either for a thorough melt or to fill the arc path, so penetration again takes a hit and undercut can be a problem.

- If in doubt and where set output, shield gas and welding position allow, use spray transfer on the thickest wire the set allows.

- Where spray isn't an option, weld the joint vertically upwards with dip transfer. The slower travel helps heat build up, and gravity drains the pool so the arc can puts its energy into the parent plates.

- **IF NONE OF THE ABOVE ARE POSSIBLE, DON'T USE MIG.** Switch to stick electrodes or take the job to someone with a more capable set. *Sudden failure of a safety-related item can be terminal.*

Table 4.1: Wire Sizes, Currents and Voltages

Wire diameter	Current (A)	Voltage (V)
0.023" (0.6mm)	50-100	16-20
0.030" (0.8mm)	60-185	18-24
0.035" (1.0mm)	80-300	18-32
0.045" (1.2mm)	120-380	18-32
Typical operating ranges		
Dip transfer	16-20V	
Globular transfer	22-28V	
Spray transfer	24-32V	

Note: *Values will vary with equipment, shield gas and welding position.*

MIG'S Various Variables

That's adjusted the equipment; now to do the same for the operator.

Which variables affect weld strength? To three of MMA's originals (current, speed of travel and angle), MIG adds voltage and wire extension.

One of MIG's selling points is that it's easy to make a reasonable job with very little practice. And it is, because the very thing that most people find hardest with MMA — holding arc length steady — is controlled by the set. But for consistent, strong welds all the various variables still have to be right.

Arc Length. The more volts or less wire speed, the longer the arc. But that's it; the operator can't adjust arc length manually during work because the set's power supply is designed to hold it constant. More voltage lengthens the arc and increases total heat input to the work, and the other way about.

Current. At a given voltage, altering wire feed speed changes current. But as we've seen, not by all that much. For large changes in penetration and fusion it's necessary to alter voltage and/or swap to a different wire diameter. Each time the operator must again fine-tune feed speed according to conditions.

Stickout or extension should be kept short, to minimize electrical resistance downstream of the contact tip and maximize welding current at the arc

(4.22). In practice use ¼"-⅜" (6-10mm) visible extension depending on welding voltage (4.23).

Welding with too much wire beyond the contact tip will hit weld strength. To see how, try this. With the arc running normally, gradually raise the gun. As stickout grows the arc gets weaker, until finally the filler wire just glows red and won't fuse with the work. By lengthening stickout you've progressively increased electrical resistance between the contact tip and arc, so energy is diverted into heating the wire (4.24).

Eventually there's not enough current left to run an arc. On top of that, increasing the stickout moves the arc farther and farther away from the gas shield, potentially adding porosity to the weld's woes. MMA electrodes don't suffer from the stickout effect. Being fatter they offer much less electrical resistance, so can carry normal welding currents without overheating.

There's a big temptation to use extra stickout when trying to weld in a deep, narrow groove or somewhere that you (or the gun) can't quite reach. But don't do it. Though the joint may seem to be welded, reduced current means that it's likely to be only stuck together.

Having said that, and as long as you stay within limits, it's OK to use stickout as a quick'n'dirty way to control current; for example on a very thin-section butt joint where burn-through looks imminent. Here a little more stickout will reduce current and prevent disaster. But don't use it as a substitute for good selection of voltage and wire speed.

Speed of travel is at the root of some big problems with MIG. Going too fast leaves a stringy bead with reduced fusion and penetration (4.25). That's

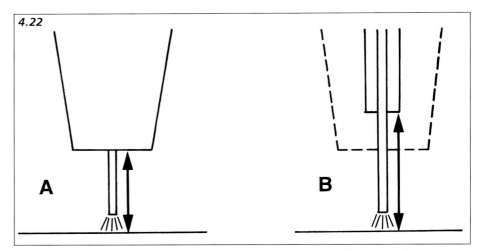

4.22. Wire stick-out or extension. Visual stick-out is what's actually seen by the operator, which is the length of filler wire outside the nozzle (A). The power supply, though, senses the whole length of resistance-creating wire between the contact tip and arc (arrow, B). The more recessed the contact tip is in the nozzle and the greater the visual stick-out used in work, the greater is this resistance. Arc energy is lost to heating the narrow wire, so fusion drops.

4.23. Generally use around ¼" (6mm) extension for dip transfer (a), rising to ⅜" (10mm) or even more where current needs to be lowered to control burn-through, but make use shield gas coverage is maintained. For the best view of what's going on, the contact tip can poke slightly beyond the nozzle if welding heat won't cook it.

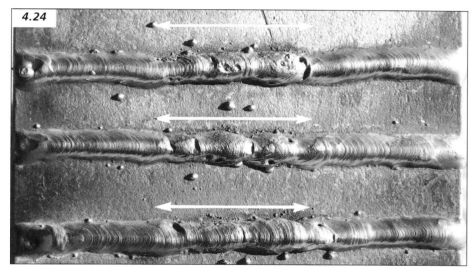

4.24. What happens when stickout gets too long? Each bead was run from right to left, starting and finishing with normal stickout. For the section between the arrows, stickout was doubled. Current dropped and the gas shield was left behind: fusion was hit, bead profile suffered and porosity and spatter grew. It doesn't take much to turn a good MIG weld into a weak one!

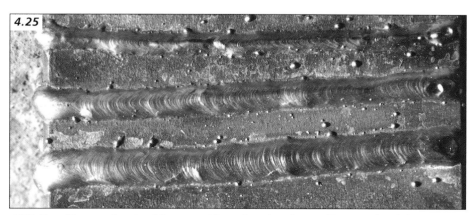

4.25. Travelling too fast (top) leaves a thin, stringy bead. Travelling too slowly leaves a heavy bead with the risk of poor fusion underneath (bottom), while travelling at a reasonable speed - which will be faster than with MMA - is just right (center).

easy to spot. Going too slowly is harder to see and if anything, is worse for strength. There's a very high risk of poor fusion and penetration, even though the weld may look fine.

Why? Rather than biting into the underlying joint metal, the arc plays on the weld pool; or worse, on a ball of molten filler rolling over the potential weld zone. Enough heat doesn't reach the joint's root and sidewalls so fusion suffers.

Here's a guide to travel speed. Once you've dialed into good welding conditions, travel at a speed that lays down a relatively flat weld with good fusion. This will be significantly faster than you're used to travelling with MMA. Watch the departing new-laid bead for height, and the joint's floor, walls and edges for fusion. Aim to keep the wire arcing at the front of a slightly sunken molten pool.

Direction of travel. With MIG there's a choice: pull or push. Normally the gun is pushed in the direction of travel at 70°-80° to the work. This is **forehand** operation (4.26a), which has a couple of advantages. If travel speed is right the filler wire feeds directly into the weld pool's forward edge, so there is less danger of the "rolling ball" effect described above. And compared to MMA, the operator gets a very good view of what's happening. In

4.26. Forehand operation is normal with MIG, that is, moving the gun in the direction of travel. This gives the wire a clear path to the front of the exposed weld pool, and the operator a good view of what's happening (a). In backhand operation (b), the gun is pulled. Backhand operation helps penetration.

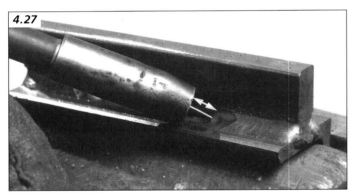

4.27. Stickout increases If gun angle gets too shallow (arrow). Current and fusion drop, and the arc tends to edge out of the shield gas envelope.

forehand operation the weld is relatively wide and flat, with reduced penetration.

Backhand operation is the alternative (4.26b). Now the gun is pulled in the direction of travel, giving deeper penetration and a narrower bead. The arc tends to be more stable and spatter is reduced.

It follows that you can, to some extent, fine-tune penetration and bead height to a given job by using either forehand or backhand travel.

Compared with MMA, MIG is more tolerant of angle variation during work. But don't use too shallow an angle (4.27). This not only reduces current through increased stickout, but also tends to carry the arc away from its gas shield.

A couple of comments on visibility and masks. When welding thin sheet with MIG, there's not much arc light about and it can be hard to see where you're going. The best solution is to splash out on a shade-adjustable, automatically darkening

4.28. A self-darkening filter (or automatic welding mask/helmet) lets you see to line up parts for tacking. Pay a little more for shade adjustability and manual/auto operation: both are well worth having.

filter. It's a small extravagance you definitely won't regret (4.28). Otherwise changing a fixed-shade filter to a relatively transparent grade 9EW will help, but be very sure to swap this for a darker item when using currents in excess of 40A: that's equivalent to 15-16V with 0.023" (0.6mm) wire. If the bank balance can stand it, keep a clearly marked mask for low-voltage MIG work and a darker one for higher voltage MIG and MMA.

MIG: Getting Down to It

After all the preamble, time to do some work! We'll start right next to godliness. **Cleanliness** in MIG work is much more critical than with MMA, as MIG wire is very short on contamination-beating ingredients. If it's true that time spent on preparation before MMA welding is never wasted, then it's doubly true for MIG. The process can handle light surface rusting and mill scale, but that's all.

The safest bet is to grind everything back to bright metal, which is often easier said than done. Where possible, get rid of all paint, significant rust, grease, mud and plating from the weld area. Where a grinder won't reach, try flame heating and a wire brush.

Most jobs can be prepared well enough. But if joint strength is critical and you can't get rid of the crud, recognize and accept MIG's limitations by switching to MMA and hoping a lot. Contamination is contamination, whatever the welding process.

Where preparation is skimpy, the MIG arc will pop and miss, sound harsh and the weld pool will "boil." Stop and you'll probably find porosity. The only thing to do is to grind out the affected section and reweld it.

You may notice a little porosity close to a start. If this is not down to contamination, it comes from a temporary shortage of shield gas. When the gun hasn't been used for several minutes, the supply pipework can empty. Give the trigger a quick squeeze

4.29. Tiny blob-tacks (a) in anything other than sheet steel won't be much good. There's little leeway for realignment (that is, belting things with a hammer) and poor fusion leaves a weak spot after welding. Tacks at least ⅜" (10mm) long are much better (b).

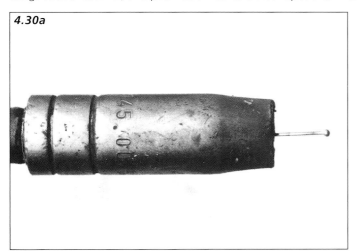

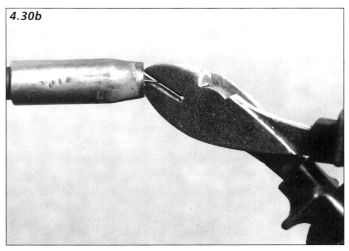

4.30. Long stickout and a burnt bulb-end on the filler wire will delay arc establishment and hurt fusion at the start of a weld run (a). For fast, clean starts and restarts, snip filler wire off at a sharp angle (b).

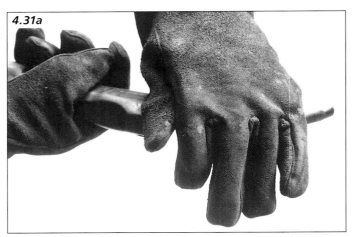

4.31. Gun control is better where two hands are used. Lay one over the swan neck (a) or rest swan neck between the first and second fingers (b).

before you start to refill the gas line, which also pushes out a little clean wire for snipping back.

MIG tacks: No differences here to MMA tacks (page 44).

Starts and stops. On non-critical work you can get away with leaving the wire's burnt bulb-end intact (4.30a), but always chop it off where high strength is important. Poor fusion at the start of a weld concentrates stress wonderfully and is a prime site for failure. A clean, fast start needs high local current which implies good electrical contact, so first see the welding return path is sound. Then snip off filler wire at an acute angle to leave a fine, sharp point; this carries high current density and promotes quick arcing (4.30b).

Snip the wire close to the contact tip to leaving minimum current-draining stickout. The shorter the initial stickout the cleaner the start, but the greater the chance of filler fusing with the contact tip as the trigger is pulled, particularly at high voltages.

It may seem a good idea to begin a run by physically touching the filler wire on the start point and pulling the trigger, but this is not good practice. As the filler wire "prop" collapses, the contact tip can touch the work and wire will want to burn back and fuse with it. If nothing else you end up wasting time

sorting out the mess. Neither is it a good idea to deal with excess stick-out by putting a 90° bend in the wire. Arc establishment will be slower as current is not concentrated into a point area, and a nasty-looking "whisker" is usually left behind.

For good control during starts use two hands. Lay the non-trigger hand over the gun's swan neck (4.31) for maximum control. Keep the gun upright to minimize stick-out and maximize current, then bring it smartly to working angle as you move off.

No matter what precautions are taken, MIG starts are prone to lack of fusion. It's a fast-moving process, so you can't wait around and weave while heat builds up as you can with MMA. If you try, the arc is likely to play on molten filler and fusion will suffer anyway. To get the best fusion at the start of a joint, use a run-on plate — a piece of scrap alongside the start area on which the arc is first established.

Restarts along an existing weld can also suffer from initial lack of fusion. To make the best job, use the MMA technique of starting just beyond the existing run's end, backtracking around the crater and continuing

on down the joint (page 42). Where appearances are important, taper-grinding the start area to a hollow pad helps. It's harder to make undetectable restarts with MIG: they don't often pass the fingernail test (page 42).

On the positive side, building-up the **end of a weld** run is easier with MIG. Just stop, circle and fill the finish area. Sometimes trapped gas erupts like a

4.32. Escaping steam or gas can crater a good finish, or heave weld metal into a solidified "pipe." Often the surrounding metal is pinholed too, so grind and refill for a high-strength job.

tiny volcano during or just after filling, threatening to leave a crack-promoting crater (4.32). This often happens when welding old pipe or box section; coating fume or water inside turns to gas or steam, whose escape route is finally blocked by the weld. Alternatively, the volcano can come from contaminants left behind after skimpy preparation. Whatever the cause, fill the crater by a short burst of weld immediately the eruption stops, or grind out the defect and reweld.

MIG Joints

Most of the basic techniques explored for MMA apply equally to MIG. But there are differences.

Butt joints. With unprepared square-edged sheet welded flat with no penetration or root gap, the upper thickness limit for full penetration is about 0.08" (2mm). Up to ⅛" (3mm) sheet can be handled by leaving a small gap, while vertical down work on anything but very thin sheet demands a gap to help penetration.

Material over ⅛" (3mm) thick will need edge preparation. A single Vee is OK up to ¹⁵⁄₃₂"-¹⁹⁄₃₂" (12-15mm), and a double Vee is best over this. If using the latter, balance out distortion by making one Vee double the depth of the other (a 30%-60% split) and fully welding the small side first (4.33).

Leaving a small step or root face at the bottom of the Vee isn't necessary with MIG, as root penetration can be more readily controlled than with MMA. If necessary, a small weave can be used on the root run to control penetration.

Horizontal/vertical joints (those where you're working side-to-side across a vertical plate) use the same attack as MMA (page 40), though now arc length is not a variable. Use the least current and highest travel speed consistent with good fusion and penetration.

T-joints are a big heat sink, which should sound warning bells if you're using dip transfer and thin wire. The weld is laid down on the edge of the

4.33. Heavy sections can be welded with MIG given sufficient heat — which means higher voltages on thicker wires. For material over ⅝" (15mm), use asymmetrical double-Vee preparation split 30%/60% as here. Weld the smaller Vee first, completing all runs before tackling the bigger part. With this technique, distortion is largely self-correcting.

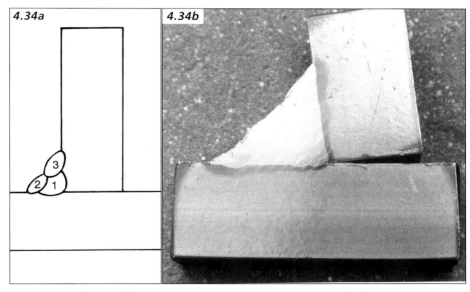

4.34. Fillet multi-runs (a) are sequenced the same as in MMA work, that is, a root pass, then a run half-overlapping the lower plate and another half-overlapping the vertical plate, and so on. Many passes went into this h/v fillet in 1" (25mm) steel (b), with no allowance made for distortion. See what the cooling weld metal has done for vertical plate alignment?

vertical plate but usually away from the edge of the horizontal one, so in total there's more metal to absorb weld heat. Thus higher energy input is called for; one voltage step up over the equivalent-thickness flat butt is normally enough, with wire speed balanced to suit. If you're using spray transfer, weld a fillet

in the flat position so it forms a Vee under the gun. The pool is hot and fluid, and away from the flat position there's always the risk of sagging and undercut.

Sequence multi-run fillet passes as you would for MMA. That's a root run first, then a half-overlap on the bottom plate and half-overlap on the top and so on

4.35. *Vertical down work with MAG is never strong, thanks to poor penetration and fusion in anything but thin sheet. But it has its uses: this long open corner joint in ½" (12mm) plate had a small gap at the top and a ¹⁹⁄₃₂" (15mm) gap at the bottom. It was welded in one pass, first with no weave and then with the gradual introduction of an inverted C-pattern to bridge the increasing gap.*

(4.34). Lower the gun angle slightly to fire weld metal up on the vertical plate in run 3, keeping an eye open for undercut.

Don't slow down to try to fill a fillet, or poor fusion will likely be the result. And for the main passes don't weave if you can help it, as fillet legs are likely to end up uneven and the greater heat input risks undercut.

Lap joints are only a different form of fillet, so the same rules apply. Use a slightly higher voltage than for the equivalent-thickness flat butt, don't work too slowly and avoid over-burning the top edge.

Vertical welding is generally easier than with MMA, as there's no slag to complicate matters (4.35, 4.36, 4.37). General points first: the weld pool is easier to control if current is kept low, so work in the lower end of the "good welding condition" wire speed range.

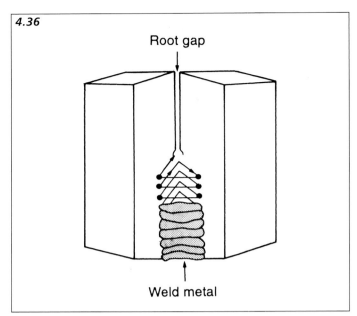

4.36. *Welding vertically up into a Vee, good penetration comes from a triangular weave (top). The arc is taken into the root with a positive pushing movement, shown by the arrows. Move the gun rather than letting stickout grow. Work deliberately, pausing at the extremities to allow filling (see dots) and not travelling too far upwards for every traverse. In tight Vees, the nozzle can restrict access. If it looks as though you won't be able to reach the joint root without excessive stickout, use MMA.*

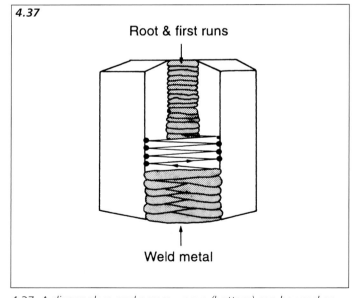

4.37. *A diagonal up-and-across weave (bottom) can be used on flatter faces, either to cover more width or to give a better-looking finish than an H-pattern movement. A C-pattern, inverted or not, gives a neat finish, too. Whichever you opt for, don't rush. Work deliberately, pausing at the black circles. Where vertical work on anything other than thin sheet is called for, travel upwards. Done properly, this guarantees penetration and strength.*

Gun angles are much the same as with MMA but are less critical. Start with the nozzle at or just above horizontal when viewed from the side, and centered over the joint line when viewed from the front. Angles can be altered during the run to control fusion and penetration. Dropping the gun angle increases stickout, lowers current and makes the pool more controllable. But don't overdo it, or fusion will fall right away and there's a risk that the arc will move outside the gas shield.

In general, keep the gun centered over the joint line to balance heat input into both plates. The obvious exception here is when welding thin material to thicker stuff, when heat can be biased towards the thick plate for better fusion and control.

Vertical down work has limited application with MIG. Penetration is never better than poor, so the technique is best restricted to sheet under ⅛" (3mm) or for gap-filling. On plate thicker than this there will always be a major lack of penetration and fusion.

Butt joints welded vertically down will always need a penetration (root) gap unless material is very thin indeed, and vertical down should not be used for anything other than an un-weaved root run or for filling gaps.

Gaps? No shortage of these in agricultural repair work. Here vertical down's lack of penetration can be turned to advantage, allowing very wide holes to be filled without burn-through. Use a side-to-side weave to bridge whoppers (4.35).

Vertical up welds are potentially very strong. Control is easier than with MMA, so going upwards shouldn't cause too much alarm. You'll generally have to weave so there's the potential for good heat build-up and fusion. Penetration

is also excellent as molten metal drains back from the weld pool, leaving clear parent plate for the arc to work on.

For vertical-up fillets, open corners and Vees, a triangular weave is good (4.36). Be very sure to push the whole nozzle into the root each time. If you let stickout grow to take the wire to the work, then fusion will suffer. As with MMA move deliberately during all weaves, pausing at the outer corners for good infill. Don't move up too far for every traverse; aim for a compact series of weld steps. Keep the arc moving relatively quickly across the weld face, or you'll end up with a series of candlewax blobs rather than a set of neat ladder rungs.

Subsequent run(s) can be made using a side-to-side weave, travelling only slightly upwards with each traverse, scudding faster across the face and pausing at the edges (4.37). Alternatively an inverted C-shaped weave works well, but watch for changes in stickout.

Vertical up's excellent penetration sets a lower limit to material thickness and penetration gap size. Sheet under ⅛" (3mm) is prone to undercut and eventual burn-through, while working in a gap means difficulty in controlling penetration and maintaining a continuous arc. Wire will shoot through, interrupting the arc and leaving an untidy "whisker" on the far side.

Overhead work. Some say this isn't recommended with MIG, but using dip transfer it's quite feasible. Use similar settings to equivalent flat welds (perhaps with a little less wire speed), and travel a little faster to avoid excess weld metal build-up. Both straight root runs and subsequent weaves are possible. Keep a closer eye than normal on the nozzle as falling spatter blocks it very quickly, especially at higher voltages. And don't forget your fireproof hat.

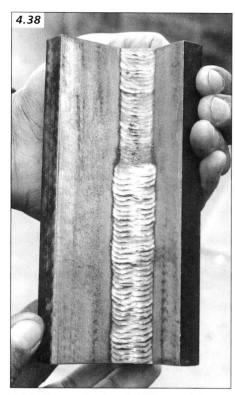

4.38. Pictures 4.36 and 4.37 turned into metal, using three vertical-up passes in ½" (12mm) plate. The triangular-weave root run can't be seen but the following two can, and they're examples of good control. The second run was another triangular weave, while the third's necessary width came from a diagonal up-and-across movement. Heat input was controlled by using relatively low voltage and wire speed, working in the lower part of the "good welding condition" range.

Faults, Problems, and Remedies

Mistakes happen. The following sections cover technique, equipment setting, equipment faults and oddities.

Technique-Based Problems

Fault: Travelling too fast. Weld bead is "stringy" and irregular; ripples are pulled into Vee-shape. Fusion/penetration is poor. **Remedy:** Slow down! (See page 63.)

Fault: Travelling too slowly. Bead heavy and wide, ripples U-shaped. Risk of poor penetration and fusion. **Remedy:** Speed up! (See page 63.)

Fault: Stickout too long. Bead irregular, fusion patchy, much spatter. **Remedy:** Reduce stickout to ¼"-⅜" (6-10mm). (See page 62.)

Fault: Weld start poorly fused or not continuous with existing weld metal (4.39). **Remedies:** Check return lead connection (high resistance lowers current and hence hurts fusion). Snip back old filler wire at acute angle to ¼" (6mm) stickout before starting. Wait for arc to establish before moving off, but don't overdo it. If necessary, use run-on plate alongside weld area to establish arc and move on to joint.

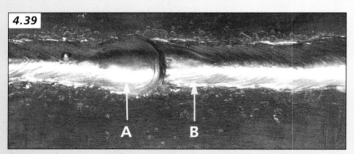

4.39. A poor MIG restart. New metal (A) hasn't overlapped the old finish (B), leaving a weak spot.

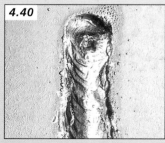

4.40. As with MMA, too much heat and poor technique makes a mess of vertical up work. But see how molten metal has drained back from the weld pool (top), giving great potential for penetration?

Fault: Severe undercut and sagging in vertical up work (4.40). **Remedy:** Lower voltage to bring heat down, reduce filler wire volume by setting wire speed to low end of acceptable range. When weaving, travel smartly across weld face to avoid heavy build-up. Angle each traverse slightly upward to

limit metal deposited, but still pause at edges to fill joint. Picture 4.41 shows several faults. The weld wanders off-line. Its lower section (1) has too much upward movement for every traverse, giving weak, big-step appearance and heavy metal deposit on face. The middle section (2) is better as upward travel for each traverse is reduced. The upper section (3) is about right, with little upward movement for each traverse and quick passage across the face. The outcome is a compact, fast weld with controlled metal deposit.

Fault: Poor joint area preparation. Result: Arc won't run smoothly and develops "machine-gun" staccato crackle; stubbing (page 59) may occur. Problem may clear itself along joint. Weld quality uneven with areas of pinholing. Joint strength poor. **Cause:** Patchy contamination along weld zone, typical of rust and oil. **Remedy:** Grind out existing poor weld, thoroughly grind unwelded area and try again. Flame heat and wire brush areas where a grinder can't reach. If good preparation is not possible, don't use MIG where high strength is important.

Equipment Setting Problems

Fault: Voltage too low. Bead is narrow, round or oval, humped on plate with very little edge fusion. Penetration and joint strength poor. **Remedies:** Increase voltage (optimizing wire speed each time) until bead flattens and fuses. If only a small increase in fusion is required, supply more current by increasing wire feed speed. Upper wire speed limit is felt by onset of stubbing. Fine heat input adjustment is possible by decreasing stickout, thus increasing current. NB: A wide humped bead can come from too-high wire speed with correct voltage. To pinpoint this, reduce speed step by step. If bead flattens and fuses, voltage is okay.

Fault: Voltage too high. Penetration/fusion excessive, leading to undercut. Burn-through very likely on thin sheet. Bead flat, may be irregular, looks "burnt" with excessive spatter. Ripples pulled into exaggerated Vee-shape. **Remedies:** Reduce voltage and re-optimize wire speed. If necessary/possible, go down to next wire size. For small heat input changes, reduce current by first reducing wire speed and then increasing stickout to maximum of ⅜" (10mm).

Fault: Weld porous. Weld metal coarse, like mousse or Aero chocolate. **Cause 1:** Lack of shield gas. **Remedy:** Is there gas in the cylinder? Is its spindle valve open and regulator set? Otherwise check gun and nozzle for spatter blockage.

If all is OK follow gas path through set, checking for flow after solenoid valve and before gun. **Cause 2:** Contamination along weld line. Grind out and reweld. If good cleanup not possible, use MMA. **Cause 3:** Filler wire rusty. Strip wire off spool until clean coils reached, then add spring clip/steel wool over wire, followed by clip/ cloth pad to clean wire before feed rollers. Keep spool in dry place when not used.

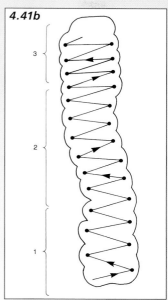

4.41. Common vertical up weave problems (a) and their breakdown (b). Consistent weaving gives a good, strong result (c).

Fault: Gun jumps in hand, smooth arc noise breaks up while stickout visibly lengthens/shortens. Cause 1: Stubbing. Wire is being fed faster than it burns back, thus physically pushing gun away from weld area. **Cause 2:** Wire speed too high for voltage. **Remedy:** Drop speed.

Equipment Troubles

Fault: Arc unsteady. Cause 1: Torch guide tube coiled or bent at sharp angle, especially where high wire speed is needed. Straighten and retry. **Cause 2:** Worn, damaged or loose contact tip. Check condition. **Cause 3:** Weld zone contamination. Thoroughly regrind unwelded area, grind defective weld and redo. **Cause 4:** Wire feed rollers slipping. Check roller pressure, spool drag setting and wire condition. Wire should be rust- and oil-free; spool drag set to just stop overrun when trigger is released, and roller pressure set so wire just stops when braked between fingers. **Cause 5:** Dirty torch liner. Blow out with dry compressed air; replace altogether if necessary. **Cause 6:** Wire drive motor or speed control circuitry unstable. Speak to machine supplier for fix.

Fault: No wire feed or wire speed uncontrollable. If motor not running with trigger pressed, check mains power and set are on (!), that speed and voltage are selected on front panel, that fuses on mains supply and any in set are OK. Check gun trigger contacts or microswitch. If motor operates but there's no feed, check roller setting and wire spool for tangles. If feed speed uncontrollable, control circuitry has gone berserk. Call the set's maker.

Oddities

Fault: Joint failure from poor fusion. Cause: low heat input. Voltage/wire speed too low, forward speed too slow, combination of thin filler wire and dip transfer. **Remedies:** Change to thicker wire/higher current if available. Use spray transfer if set capacity and wire diameter allow for flat work and horizontal-vertical fillets. Weld vertically upwards in dip transfer mode to increase heat build-up. Travel at speed that keeps wire arcing to front of visibly fusing molten pool. If extra heat input not possible through set output/ positional limitations, switch to MMA.

Fault: Volcano-like crater or pipe-like knob at end of run. Snapping off projection shows pinholed weld metal underneath. Caused by contamination from escaping gas or steam. **Remedy:** Fill crater with burst of weld while still red hot, grind off projection and reweld.

Fault: Arc wanders uncontrollably between the wire and work. Cause: Arc blow, typical of DC processes. Strong, directional magnetic fields are pulling the arc out of shape. Typically happens working in a tight corner or when holding work with a magnetic clamp. **Remedy:** not always 100% curable. Keep the welding return lead's clamp close to the weld area. Weld directly toward or away from it. Swap magnetic clamp for mechanical one or move it further from weld area.

Gasless MIG

Having discovered the simple joy of MIG welding, you'll find it stops once you wheel the set out of the workshop. Unless the weather is very quiet or you've cobbled together a shelter, the wind will whip away the shield gas to leave holey (and wholly) unsatisfactory joints. And that's a bummer.

But what if the welding wire could pretend to be a stick electrode and take along its own gas shied? Cored MIG wire does just that, as farms using it for hardfacing will already know. Rather than being solid like the usual stuff, cored wire is a hollow tube packed with flux . . . like a stick electrode turned inside-out. And it's usable with conventional sets, given several limitations and some changes in setup/technique.

The box "Roll Your Own" details how cored wire is made. Like most politicians it's been called several names; flux-cored, gasless or self-shielded wire are the most common. Developed first in the 1950s for shipyards and other heavy industries that can't wait until the breeze stops blowing, cored wire can join most materials and do pretty much everything that a stick electrode can, including handling more contamination than solid MIG wire.

But don't shout "Eureka!" and throw your hat in the air. Most cored welding wires are designed for DC negative sets, while the smaller single-phase units found on farms are universally DC positive. On top of this, cored wire is expensive: a kilo costs around five times more than solid wire. So while it can bring some of MIG's speed and ease of use to the great outdoors, cored wire is far from a magic bullet.

WITCHES' BREW

Much of cored wire's ability comes from the flux, whose main jobs are to shield the molten weld from oxygen and nitrogen, stabilize the arc, squirrel away impurities in the slag and then protect the cooling weld.

Sound familiar? It should — the core powder is much like the flux on a stick electrode, only not bound into a solid mass. De-oxidizers and de-nitrifiers handle contamination from the air, other components help wetting and cleaning, while metals like chromium and molybdenum can go in to boost weld toughness.

Being packed with flux, core wires are bulkier that their solid equivalent. The smallest diameter on offer is 0.035" (1.0mm) (against 0.023" [0.6mm] in solid wire), with sizes more normally spanning ¹⁄₁₆"-⁵⁄₃₂" (1.6mm-2.4mm). This affects the minimum welding current required and the minimum material thickness that can be joined, as we'll see later.

Cored wire comes in many grades. But many won't work without a separate booster supply of shielding, usually CO_2 or an argon/CO_2 mix from a cylinder. Of those that can go solo most need to run DC negative; that is, with the torch plugged into the negative side of the power supply. Which as we've said, run-of-the-mill MIG sets can't handle.

Cross off wire types that won't work in farming and the choice is cut dramatically. For example ESAB offers one wire — OK Tubrod 14.18, with equivalents on offer from other brands. It's is a very simple beast, able to run DC positive and joint mild, medium-tensile and cast steels in a moderate breeze. Comparable to basic ER70S-6 MIG wire and more versatile than a general-purpose mild steel rod, it's close in nature to an E7016 low hydrogen stick electrode. The snag is that this wire is not happy on vertical or overhead joints, which limits usefulness for stockyard and machinery repair.

SETUP AND TECHNIQUE

Three areas change on switching to cored wire — the torch, welding technique and plant settings. We'll start at the business end and work out from there.

• **The torch.** With no shield gas there's no need for a nozzle, so off it can come. This leaves the torch tip exposed like a skinned rabbit. Access into tight places is much better, but you'll need to watch tip positioning (as of course it's live when the arc is running) and take care not to ding any torch threads left exposed. If you're feeling flush, buy a specific cored wire torch. The set's existing feed rolls and torch liner can probably be used as they stand, remembering to change rolls to suit the new wire diameter. If these can't deliver an even feed without major pressure increase, replacement knurled rollers should do the business without crushing the wire.

• **Technique.** Here as always preparation comes first. Thanks to the flux's scouring abilities, cored wire can handle as much contamination as a good mild steel stick electrode. This makes it particularly good for joints where poor access renders 100% cleaning impossible. Grind off paint and platings where you can and don't skimp, but cored wire can put up with modest rust and mill scale if it must.

• Next comes **travel direction.** Cored wires work best when the torch is pulled, rather than pushed. That is the tip trails the operator's hand, as with stick welding. Pulling the torch keeps molten weld metal more where you want it, maintains the best gas cover and produces a fuller, more even slag coat on the weld. Travel speed is set by the same considerations as for all MIG: keep the arc concentrated more on parent metal than the molten pool, and look to fill the joint and fuse the edges.

• **Stickout.** See page 62 for definition. Stickout must change dramatically. The flux in a cored wire needs to be

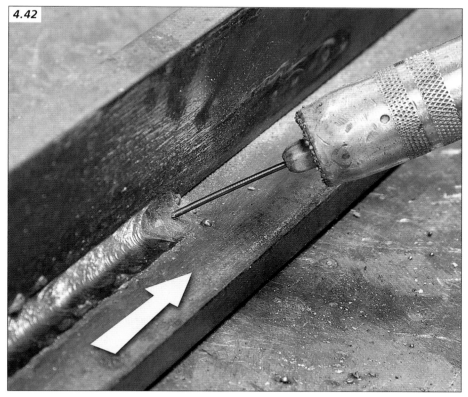

4.42. *Cored wire brings changes from conventional MIG technique. The torch is best pulled; that is when the welder's hand leads the arc, as here. And visual wire stickout (the distance between the contact tip and arc) is much longer: 1"-1³⁄₁₆" (25-30mm) rather than the usual ¼"-⅜" (6-10mm). This torch still wears its nozzle, though shield gas isn't flowing. Take it off for better access into tight spots.*

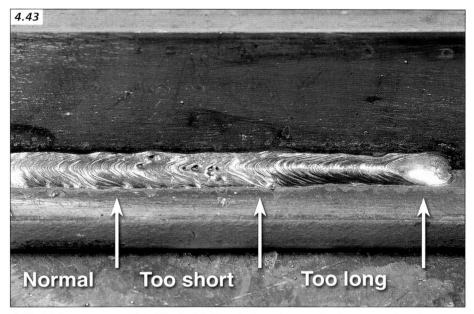

Normal   **Too short**   **Too long**

4.43. *Stickout matters. This fillet in ¼" (6mm) plate shows what happens when it strays too far. Holding stickout at 1"-1¼" (25mm-30mm) allows good flux pre-heating and produces a tidy result (left). Halving stickout cuts flux effectiveness and leaves a weak pinholed weld (center). Extending stickout too far drops arc current so there's not enough heat to burn off the wire: the bead is humped with poor fusion and penetration (right).*

warmed before it reaches the arc or it won't do its best. The more stickout you give, the hotter flux gets through resistance heating of the wire. Try 1"-1¼" (25mm-30mm) as a starting point (4.42). Here's yet another of welding's balancing acts. Too little stickout stops the flux working properly, so gas cover breaks down and you get pinholes in the weld. Too much stickout bleeds off energy into heating the wire, so there's less to run the arc and penetration/fusion suffers. Take a look at 4.43 to see all this in action.

• **Voltage and wire speed.** Compared with a solid wire, the cored version's thin walls have a much smaller cross-section to carry current. Lower voltages and wire speeds are needed, yet the high current concentration still gives good potential penetration. On the other hand, burning off the flux takes substantial energy. So the welding set has to be able to put out around 200A minimum for ¹⁄₁₆" (1.6mm) cored wire.

As with normal MIG operation, each voltage step on the set links to a corresponding band of feed speed to deliver good welding conditions. In this band voltage and wire feed are in balance. Expect to use lower speeds and voltages than with solid wire, and as usual your ears and eyes confirm when you've hit the sweet spot — the arc pops and crackles, albeit more softly than with solid wire, and spatter is at a minimum.

Cored wire should feed into the weld pool steadily without whooshing or stubbing, though there's usually a dewdrop or "boot" of molten material at the tip. Compared with conventional dip-transfer MIG work there will be more spatter as some metal is transferred across the arc in globular way; but otherwise the joint should be neat and tidy (4.44). Slag is thinner than with stick electrodes, lifting off easily (4.45). Chip and wire brush between runs. Slag traps are less likely than with stick electrodes if voltage and wire speed are right.

OTHER LIMITATIONS

Minimum diameter of ESAB's Tubrod 14.18 wire is 0.035" (1.6mm), so it needs a fair whack of current to run properly. This calls for a capable welding plant. It also defines the minimum material thickness weldable without burn-through — around ⅛" (3mm). Having said that, the process's high tolerance to variable stickout length and freedom from a separate shield gas blanket mean that thinner stuff and irregular gaps can be tackled, adjusting welding current on the hoof by varying stickout. Heavy sections can be welded with appropriate preparation and multiple passes. No special pre-heating or cooling is needed with mild steel.

The last note is a caution. Welding with cored wire produces a lot of fume, which is definitely not good stuff to breathe. Outdoor work on a breezy day should be fine, and that anyway is the very time when you'll want to exploit cored wire. Indoors, weld as close as possible to an open door or in moving air from a window. Ideally use a fume extractor. If that's not possible, wear a disposable respirator under your welding helmet or invest in an aspirated filter set.

4.44. Cored wire MIG produces smart beads marked out by spatter. The latter is inevitable as metal transfer is partly by dip and partly across the arc, so some gets lost on the way.

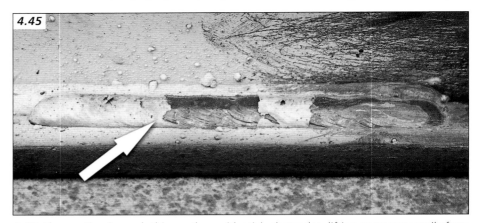

4.45. Flux coating is much thinner than with stick electrodes, lifting away very easily from a tidy bead.

Roll Your Own

How do you make welding wire that thinks it's a sausage roll? By sneaking up on it in small steps. Steel arrives at the factory as rolls of flat sheet. This is cut into smaller and smaller strips by drawing it through roller guillotines. Once the strip is the right width a series of shaped rollers form it into a U-tube, which then passes at right angles just below a wide conveyor belt.

Down the belt comes a stream of powdered flux, spread in a thin layer the width of the belt. Flux rains in a steady curtain on to the U-tube whizzing by below, partly filling it. More rollers then curl the tube closed.

By now the wire is around 0.2" (5mm) in diameter. To bring it down to the final size (0.035"–³⁄₃₂" [1.0mm–2.4mm]) it's drawn progressively through several sizing dies, compressing the flux core and getting mighty warm along the way. All this happens on a long, humming machinery line. Finally, the sized wire is wound on to big spools before taking the short hop across to the spooling area, where it's dispensed either on to consumer-sized reels or big 550 lb (250kg) packs for use in robot welders.

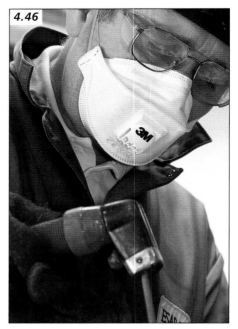

4.46. A disposable respirator.

Section 5

Gas Welding and Cutting

As MIG sets come down in price and move into farming, buyers seem less interested in gas welding. This is a real pity. For after a credit card, oxyacetylene actually is your "flexible friend." Nothing else can fusion weld, bronze weld, braze, cut and heat. Even if gas gear's welding ability is ignored, having a flame for bending, straightening, cutting, unseizing bolts and creating general mayhem is more than just convenient. In any serious workshop it's essential.

Buying: New or Secondhand?

Unless you're quite sure what you're looking for and at, buying secondhand gas equipment is not a good idea. With gas equipment it's quite possible to blow yourself up, and using worn-out gear is an easy way to go about it. So tread warily round a farm-sale set or local-paper bargain. With any gas gear the motto has to be "If in doubt, don't buy." Let someone else have the bonfire: safety must come first.

If you're still tempted, look for damaged fittings; for cracked, bent or otherwise knocked-about gauges; for damaged seats on all hardware, and for split or cracked hoses. Some sizes of oxygen cylinders from are now filled to 3,450psi (230 bar) pressure, so it's essential to be sure that a used regulator is marked either "3,450psi service" or "4,350psi service" ("230 bar service" or "300 bar service"). Older regulators will only be rated to 3,000psi (200bar) and must not be used where there is the slightest risk of connection to higher-pressure cylinders.

Be aware that gas cylinders remain the property of the supplying company so won't be the vendor's to sell. Before equipment changes hands, the existing cylinders should go back to the nearest depot and any outstanding rental should be settled. The remaining hardware can then be sold, with the buyer renting fresh cylinders in his/her name.

The best way to own gas gear is to buy new. Looked after, it'll last a very long while. Work out your requirement in terms of material thickness for welding and cutting, then trundle off to a reputable supplier. He'll fit you up with quality kit and be there when service is needed.

Backfires, Flashbacks, and Arrestors

You'll hear occasional pops and bangs during welding or cutting. At best the pop makes you jump; at worst you're on the way to trouble. What's happening? Small-scale bangs are **backfires**. Gas is igniting spontaneously, rather than burning steadily at the nozzle. Common backfire causes are:

• The nozzle touching the work or temporarily blocking with spatter. Gas flow stops, the flame goes out and then reignites with a bang. If this blows the

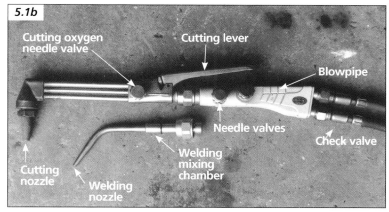

5.1. Equipment for oxyacetylene welding and cutting. Note the flashback arrestors in (a). These are essential, both for secondary safety and to fulfill a farm's duty of care to employees.

The Big Bang . . . Please Don't Have One

Picture 5.1 covers naming of parts. As it's wholly possible to do yourself and others mortal damage with oxyacetylene equipment, no excuse is offered for devoting space to the ground rules.

Please don't flick through this section. Understand exactly what's what before trying out new or secondhand gear. If in doubt over any procedure, ask the gas or equipment supplier for advice.

Gas cylinders are best treated as potential bombs. A cylinder explosion in a workshop doesn't leave many pieces to be picked up, and you get no second chance. The cylinder supplier is required by law to inspect and pressure-test each unit regularly, so although a fresh cylinder may have leaked away some of its contents before coming home, it's intrinsically safe. Explosions and fires are thus down to bad practice or equipment neglect. Safety is solely the operator's business.

So how best to look after the kit and (by implication) yourself? The rules aren't hard, though they may seem many.

Cylinder care is a good a place to start. Acetylene gas is only slightly compressible before it detonates, so it's sold dissolved in acetone. The acetone is held in an inert matrix but isn't fixed by it, so laying a cylinder down lets liquid collect in the neck. If this acetone is allowed to reach regulator diaphragms and hoses, it attacks them with eventual safety consequences.

Thus cylinders must be used upright, and ideally moved and stored upright. But it's not always practical to move a cylinder without laying it down. To let acetone drain back where it belongs, a cylinder that's been horizontal for more than a few minutes should be left vertical for at least a couple of hours before use, and eight or more hours is better. And yes, people do get away without doing this. But acetone gnaws away at rubber parts over time, unseen. Eventually equipment will fail, always with expensive (and possibly with nasty) consequences.

Regulators and hoses don't last forever. In the UK, regulators must carry a date of manufacture. Because their internal seals and diaphragm degrade over time, the British Compressed Gases Association's code of practice (CP7) recommends regulator replacement every five years. Although not law itself, CP7 lays out the minimum needed to meet duties under law regarding gas equipment, and recommends periods for equipment inspection as well as replacement.

Fitting regulators to cylinders is straightforward. New cylinders arrive with plastic protectors on their outlets, so these must come off. Regulators and fittings for the fuel gases acetylene and propane have left-hand threads, while oxygen threads are right-handed. Thus the two can't be mixed up. But don't be tempted to swap fittings between different fuel gases:

5.2. *Portability comes from small cylinders on a trolley which accept normal peripherals. Cylinder volume is OK for welding, but cutting soon empties the oxygen.*

the threads may allow it, but gases vary in delivery pressures and permissible flow rates. Using inappropriate bits can end in tears.

With regulators to hand that matches both the gases and maximum cylinder pressures, clear the cylinder necks of water and insect life by a quick "sniff" — that is, crack open the spindle valve a little so gas can blast out debris. Then fit the regulators so that their dial faces can be easily seen and their hose outlets don't interfere with spindle key action. Tighten regulators and hose connections with the wrench supplied. If this has long since gone into a tractor driver's toolbox, use a normal-length wrench. *If any joint leaks after reasonable tightening, find out why rather than overtightening or trying to repair leaks with tape.*

Regulator life will be longer if the central pressure-adjusting screw is backed off after use. Not between jobs, but overnight or through the weekend. If left screwed in, the underlying diaphragm and spring will slowly take on a permanent "set" and accurate adjustment eventually won't be possible. There's a safety angle, too. With the regulator screw backed right off, gas can't escape down the hoses when forgetfulness or a leaky spindle valve leaves gas in the regulator.

Oxygen regulator gauges carry the warning "Use no oil." This is not whimsy: ignore it at your peril. High-pressure oxygen meeting oil or grease anywhere in the system (including the blowpipe) can give a tidy explosion. If you must lubricate sticky fittings, use only dry graphite.

Should an oxygen cylinder regulator blow or be sheared off, it's a bit like letting go of the neck of a huge balloon. But this balloon is made of steel, and nothing will stand in its way; not walls, not equipment and certainly not a human body. For this reason it's major folly to carry or lift a cylinder by foreloader or forklift, and hoisting with slippery chains or weak rope is right out. Keep your cylinders (including spares) chained into a trolley or tied back against a wall.

More Shock Horror

We haven't finished yet. Clumsily repairing hose connections can lead to trouble through leaks, as can using old, perished hoses. Keep hoses looped in gentle bends while out of use, as wrapping them tightly accelerates cracking.

Don't be tempted to repair a hose with a length of copper pipe, as copper reacts with acetylene to produce unstable, explosive acetylide compounds.

Check hose and blowpipe connections now and again for tightness. Acetylene's distinctive whiff is easily picked up, so nose round the regulator and pipework; the gas smells like someone's breath the morning after a particularly significant curry. Only a true dumbo would use a match to find an acetylene leak, and then they'd only do it once.

Oxygen has no smell, so track down leaks by ear or by using a dilute soap solution. Brush it on and watch for bubbles. Leaks from spindle or needle valves can usually be cured by controlled tightening of the gland nut, but don't overdo it.

Though not inflammable itself, oxygen very actively supports burning. Oxygen-impregnated clothing will flare like a torch if a spark lands on it, and people have died this way. *It follows that oxygen must never be used in place of compressed air to blow down clothing, to clear a bench or to power air tools.*

Finally, for anything short of very heavy cutting, don't open either cylinder valve more than one full turn or it can't be shut quickly in an emergency. Less is usually plenty for normal gas delivery. And always park the spindle key on the acetylene cylinder's valve, for this is the one that must be shut first in case of trouble.

That's a run-through of the main safety points. Others will be covered as they come up.

---

flame out, don't try relighting with both gases flowing or another, possibly bigger bang may result. Shut off acetylene first at the blowpipe needle valve, then oxygen. Purge both gas lines and relight (page 78).

• The nozzle overheating through working in a confined space, so the gas mix fires off inside it. To cool it down, turn off the acetylene flow and plunge the whole nozzle into warm water: a cold bath can cause cracks. Oxygen flow will keep water out while the unit is pretending to be a submarine. Check nozzle tightness afterwards as thermal shock can loosen it.

• A worn nozzle. When the gas exit hole(s) wear to a bell-end, backfire risk goes up. Reface or replace the tip (page 90).

• Low gas pressures. If the gas column is moving out of the nozzle too slowly, the flame will burn back down it. In marginal pressure conditions the result is isolated "pops" or sometimes a machine-gun string of mini-explosions.

In worse cases this popping escalates into a sustained backfire. The flame disappears down into the blowpipe's mixing chamber and burns there; you'll hear a thin screeching noise and see a stream of black smoke coming from the nozzle. Act quickly now, or the blowpipe can melt. **Shut off oxygen to starve the flame** (the only time when oxygen is shut down first) and then the fuel gas. Purge both lines (see page 78) before relighting. To stop the banging increase pressures a little, keeping them balanced if appropriate.

Don't let backfiring go on as it may lead to worse things. Normally, gas is fed down two hoses into the blowpipe. Where feed pressures are not exactly equal (and they seldom are), gas from the higher pressure line wants to push into the lower pressure one. If it can do so the result is a potentially explosive mix in one hose, which could be triggered by a backfire's flame front. To head that off, OSHA requires that check valves (5.3) are fitted at the blowpipe end of each hose. These are one-way

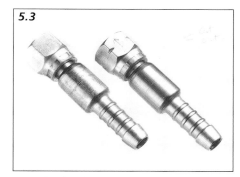

5.3. Check valves are a legal requirement at the blowpipe end of hoses. Test often by blowing through them — an arrow (top right) shows direction of gas flow.

valves which close when gas tries to backtrack. A backfire's relatively low-pressure, cool flame front is stopped by check valves.

Test the valves occasionally as carbon build-up can turn them leaky. Unscrew the hose, put the valve end under water and blow down the other. If the check valve is OK, no air should pass in the regulator direction but flow should be free towards the blowpipe.

Check valves are only a first line of defense which can't handle all eventualities. It's possible in normal use to generate extreme pressure imbalance; perhaps when something steps, falls or drives onto a hose, when it kinks, or when one cylinder runs empty of gas. At this point you can be heading towards a flashback.

With one hose blocked or out of gas, any backfire will be much more severe. If the flame is fast enough to jump the check valve(s) and enter the hose(s), these can burst. The associated high pressure wave can destroy the regulator(s) and detonate the acetylene cylinder contents, producing a lethal explosion.

So flashbacks are very bad news. Good working practice is the key to avoiding trouble, but both common sense and OSHA strongly recommends fitting **flashback arrestors** to both regulators (5.4). These snuff out a flame front before it can get to the regulator, with some designs also tripping a valve to stop gas flow. Costing from $20-$60 at the time of writing, they should be on all farm gas equipment.

Mind you, such mechanical backup should never be needed. And they won't be if you sort out the cause of backfires before they escalate, watch hoses around stock and machinery and don't run cylinders right down to their last dregs. *And before lighting or relighting a blowpipe, you purge each line completely to make sure it contains only the gas it should.*

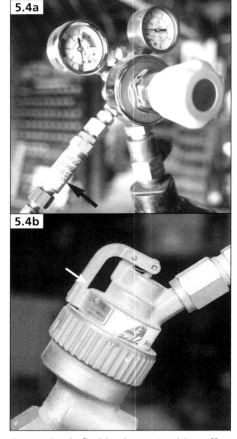

5.4. A simple flashback arrestor (a) snuffs out a high-velocity flame front before it can reach a cylinder. Some also have an internal link that melts in flame heat to cut off gas flow, while others can reset automatically. More expensive varieties (b) have pressure-sensitive valving and are resettable via an external arm (arrow).

Table 5.1: Safe start-up/shutdown procedure for oxyacetylene welding equipment.

1. Make sure blowpipe needle valves are closed. Back off regulator screws if necessary. Check all connections are tight, especially at the blowpipe.
2. Stand facing regulator dials. These usually have blow-out backs, so facing the dial is safest. Open acetylene cylinder spindle valve slowly to a maximum of one turn; slow opening means no pressure shock to regulators. Ditto oxygen spindle valve. Leave key in acetylene spindle valve.
3. Gas is now into both regulators. Two-stage models will show pressure on one dial.
4. Purging and pressure setting are combined. Open the blowpipe acetylene needle valve. Adjust pressure regulating screw so gauge shows nominal pressure required by nozzle (table 3.3). Let gas flow for a few seconds to purge system, then close needle valve. Repeat the procedure for oxygen. NB: Pressures can be set accurately only under free-flow conditions. Back pressure from a closed needle valve gives false reading.
5. With the nozzle pointed upwards, open acetylene needle valve a little and light gas. Use spark igniter; matches or a lighter risk burning fingers.
6. Open the needle valve further until acetylene flame is 'bright' – smoke at tip is all but gone. If this can't be achieved, increase pressure at the regulator until it can. If flame separates from nozzle tip, gas speed is too high. Close needle valve until it rejoins.
7. Gently open oxygen needle valve and feed in gas until required flame type is achieved (page 80). Again, slightly increase pressure at regulator if necessary. Fine control of flame type comes only from the blowpipe needle valves.
8. To shut down, close acetylene needle valve first. Flame thus goes out quickly. Turning off oxygen first leaves a smoky acetylene flame, able to burn up to – and perhaps inside – the nozzle. The only exception is in the case of sustained backfire, where the nozzle squeals and puts out black smoke. In this case turn off oxygen first to starve the fire.
9. Close oxygen needle valve.
10. Close acetylene cylinder spindle valve, then ditto for oxygen. This traps gas inside cylinders.
11. Drain gas from rest of system by opening blowpipe acetylene needle valve. Line pressure gauge will fall to zero. Repeat for oxygen.
12. If welding has finished for more than a couple of hours, back off regulator adjusting screws. This prolongs regulator life as well as stopping the progress of any gas leaking past cylinder spindle valve. NB: Cylinder pressure gauge may creep back up over time if spindle valve is leaking. Don't be tempted to jam valve shut by hammering the key or using pipe for extra leverage!

Note: Accurate flame balancing is possible only with nozzle in good physical condition.

Lighting and Shutdown: One Step at a Time

Tables 5.1 and 5.2 lay out the safe step-by-step procedure for lighting and shutting down gas equipment. Follow it to be sure that gas flow through equipment is under control and that lines contain only one gas. **Purging** means clearing the system of any mixed gas.

GETTING STARTED

First, eye protection. The intense glare from gas welding/cutting plus the risk of physical eye damage from flying sparks makes good goggles essential, even for the briefest operation. Lenses should be made to ANSI Z71, with shade density chosen for the job. When the lenses have pitted and the strap has as much elasticity as a piece of string, give your eyes a treat and buy new goggles.

With those sorted out, the equipment safely set up and the lighting/shut down procedures understood, all that's left is to choose a nozzle, set the pressures and adjust the flame. Welding comes first, cutting on page 89.

Which nozzle to use? Table 5.3 relates material thickness to size. But experience soon shows that thickness isn't all that matters. The idea is to find a nozzle that brings work quickly up to heat, yet isn't so big that the weld pool is hard to control.

Most farms will use oxyacetylene welding on steel sheet. Work on stuff thicker than ⅛" (3mm) gets increasingly slow, takes a lot of heat, causes considerable distortion and uses more and more gas, so arc welding takes over. Most jobs will be covered with nozzle sizes 1, 2, 3, 5 and possibly 7, though one-offs may call for something bigger.

Nozzles are usually stamped with their size (1-90) and the maximum gas volume delivered per hour. So a number 1 nozzle is also stamped "26," a number 2 "57" and so on. The smallest nozzles are used where modest heat input is enough, while a number 90 will roast

Table 5.2: Safe start-up/shutdown procedure for oxyacetylene cutting equipment.

Start with all three blowpipe needle valves shut. Then follow points 1–5 in Table 3.1 to purge hoses, set initial pressures and light the acetylene flame.
6. If the acetylene flame still smokes with needle valve fully open, increase regulator pressure a little until it stops.
8. Feed in oxygen using cutting head spindle valve until flame is neutral.
9. Operate cutting lever and check flame stays neutral. If not, bring to neutrality while holding down cutting lever, using attachment's needle valve. If one or more pre-heater flames are shorter than the others, shut down. Slacken the nozzle nut and turn the nozzle so its inlet holes are not in line with those in the head. Tighten, purge and relight.
10. Shutdown is the same as table 3.1. After draining oxygen line of gas, close both oxygen needle valves.

Table 5.3: Welding nozzle sizes, recommended pressure settings and material thickness weldable. (Courtesy Murex).
See how gas pressures rise with nozzle size but stay balanced.
Figures are a guide only, and refer to clean mild steel worked indoors.

Mild Steel Thickness			Nozzle	Operating Pressure				Gas Consumption			
				Acetylene		Oxygen		Acetylene		Oxygen	
mm	in	swg	size	bar	psi	bar	psi	l/h	ft3/h	i/h	ft3/h
0.9	0.0354	20	1	0.14	2	0.14	2	28	1	28	1
1.2	0.045	18	2	0.14	2	0.14	2	57	1	57	2
2	0.0787	14	3	0.14	2	0.14	2	86	3	86	3
2.6	0.1024	12	5	0.14	2	0.14	2	140	5	140	5
3.2	1/8	10	7	0.14	2	0.14	2	200	7	200	7
4	3/32	8	10	0.21	3	0.21	3	280	10	280	10
5	3/16	6	13	0.28	4	0.28	4	370	13	370	13
6.5	1/4	3	18	0.28	4	0.28	4	520	18	520	18
8.2	5/16	0	25	0.42	6	0.42	6	710	25	710	25
10	3/8	4/0	35	0.63	9	0.63	9	1000	35	1000	35
13	1/2	7/0	45	0.35	5	0.35	5	1300	45	1300	45
19	3/4		55	0.43	6	0.43	6	1600	55	1600	55
23	1		70	0.49	7	0.49	7	2000	70	2000	70
25	1+		90	0.63	9	0.63	9	2500	90	2500	90

$psi = lb/in^2$ or lbf/in^2

Table 5.4: Nozzles and pressures for flame cleaning, acetylene fuel gas

Nozzle	Fuel gas pressure		Oxygen pressure	
	Bar	psi	Bar	psi
2" (50mm) flat	0.49	7	0.57	8
4" (100mm) flat	0.70	10	0.70	10
6" (150mm) flat	0.85	12	0.85	12

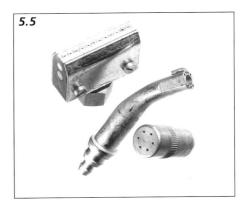

5.5

5.5. Nozzles to extend equipment usefulness. The flame cleaner (top) is good where a diffuse heat source is required, like taking off paint or surface rust from otherwise unreachable areas prior to welding. A gouging nozzle (center) cuts grooves, also useful prior to welding where a grinder can't reach. For rapid, economical heating prior to unseizing or straightening something, the heating nozzle (bottom) is best. All the above come in different sizes and may need a different mixer for the blowpipe.

5.6

5.6. Holding spare nozzles in a wood block keeps them out of harm's way — and you know where to find them.

chickens at a distance and empties cylinders in minutes.

Where thin mild steel is to be welded, start with a number 1 nozzle.

FLAME SETTING

A gas flame's nature changes with the balance of oxygen and acetylene being burnt, with different flame types suited to different jobs (5.7). Most welding needs a neutral flame; bronze welding and brazing calls for a slightly oxidizing flame, while aluminum and some hard-facing applications need a gently carburizing flame.

The three types are easily produced by twiddling the blowpipe needle valves, with differences best seen through gas goggles. Here's how they run.

Have a peek at the nozzle's tip — it needs to be clean, with a square-edges round hole. If it's worn, re-face it (pages 89-90). Set regulator pressures using Table 5.3, purge hoses and light the acetylene. Open the blowpipe needle valve until smoke at the flame tip all but disappears. If this doesn't happen, increase acetylene pressure at the regulator until the flame just separates from the nozzle tip. Then reduce pressure until the flame just sits back on the tip.

Feed in some oxygen from the needle valve. The flame turns from yellow to white. Adding a little more oxygen resolves the flame into three sections — a blue central cone next to the nozzle, a white "feather" beyond it and an outer, less distinct envelope.

Adding still more oxygen shortens the feather. All the time you can see it there's more carbon-rich acetylene in the mix than oxygen, so the flame is **carburizing**. It's also cool. Working with such a flame leaves a weld that's hard but brittle.

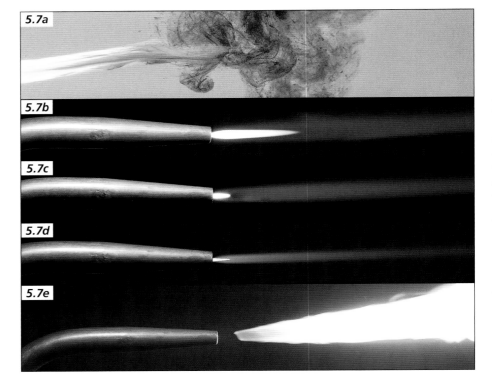

5.7a

5.7b

5.7c

5.7d

5.7e

5.7. The right flame matters. With no oxygen, acetylene burns ragged yellow with much soot and smoke at low flow rates (a). If flow rate is too high, the flame separates from the nozzle (e). Feeding in some oxygen quiets the flame and produces a long, white central feather beyond a small central cone (b): this is a **carburizing** flame. Adding more oxygen sees the feather shrinking until it's just merged with the central cone, leaving a faintly fuzzy outline (c). This **neutral** flame is used for most work. Dialing in more oxygen sees the flame shortening and getting noisier with a tiny, pointed cone (d). There's now an excess of oxygen - an **oxidizing** flame. To check for neutrality, take out oxygen via the needle valve until a feather appears, then add it until neutrality is achieved. Watch flame condition during welding as it's prone to change.

As oxygen supply is increased the feather blends down into the central cone, which turns soft-edged and hazy. Oxygen and acetylene are now being burnt in equal volumes, leaving a hotter **neutral** flame. Neither carbon nor oxygen is left over to combine with weld pool metal, so potential weld strength is good.

Feeding in yet more oxygen tips the balance again. With an excess of O^2 the cone gets smaller, harder-edged, more pointed, lighter in color and takes on a spiteful "hiss." Now it's an **oxidizing** flame, which—where not wanted—will produce showers of sparks and convert molten metal into weld-weakening oxides. If this flame can't be produced with the oxygen needle right open, gradually increase regulator pressure until it can. Then close the needle valve until the flame is neutral.

Play around with flame types so that you can produce each at will, aiming to set the regulators so that final flame tuning is done with the blowpipe needle valves. As equipment warms up a neutral flame tends to become oxidizing, which needs watching during work. Cheaper single-stage regulators can't hold pressure as constant as two-stage versions, so pressure drift can also unbalance the flame. If it changes with a regular pulsing rhythm, watch out: gas pressure is too low on one of the lines, making a check valve oscillate. Correct this before backfiring starts.

Gas Welding

With the smaller nozzles at least, welding by flame is a much quieter process than MMA's sound and fury. Part-science, part-art, it best suits men of easy temperament.

For mild steel there's a choice of processes, either fusion welding or bronze welding/brazing. In **fusion**, metals are taken to melting point and flow together in a liquid pool. If necessary, filler rod of similar composition is added to build up or fill the joint. But in **bronze welding and brazing**, the parent plates are not melted. Instead they're brought to

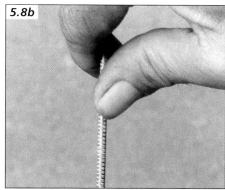

5.8. Bronze welding flux can come in a can (a) or as part of the rod (b). Here small notches in the rod's flank hold it.

red heat, then a relatively low melting point copper/zinc alloy is flowed on to the surface. This acts like a metallic "glue," hooking into tiny surface hills and valleys and locking the joint together. To promote cleanliness and surface wetting, bronze work needs a flux. This either comes from a pot, or is carried on (or in) the welding rod (5.8). Each process has its advantages, disadvantages and place.

Fusion welding needs slightly less preparation, and (given good execution) is stronger than a bronze weld. But the process puts more heat into the job, increasing distortion.

Bronze welding calls for very thorough preparation, but its lower heat input

5.9. Brazing relies on liquid bronze's ability to be sucked into a joint by capillary action.

minimizes distortion. Different thicknesses are easily joined, and (given the right rod) dissimilar metals can be welded. And it's easy to do.

Brazing uses the same fillers as bronze welding, but exploits the liquid alloy's ability to move between close-fitting plates or tubes. Capillary action sucks in the hot bronze, producing an extremely

Table 5.5: Commonly used gas welding rods for steels.
(Examples from the Murex range)

Trade Name	Suited to	Diameters (in)
Process: Fusion welding		
Saffire mild steel	Low carbon steels	1/16, 0.08, 3/32, 1/8, 3/16 (mm: 1.6, 2.0, 2.4, 3.2, 5.0)
Saffire Medium Carbon Steel	Most low-alloy steels	1/16, 3/32, 1/8, 3/16 (mm: 1.6, 2.4, 3.2, 5.0)
Process: Bronze welding and brazing		
Saffire silicon-bronze	Steel, copper, brass	1/16, 3/32, 1/8, 3/16, 1/4 (mm: 1.6, 2.4, 3.2, 5.0, 6.0)
Saffire nickle-bronze	Steel, malleable iron	1/16, 1/8 (mm: 1.6, 3.2)
Saffire manganese-bronze	Cast or malleable iron	1/8, 3/16 (mm: 3.2, 5.0)
Self-fluxing rod for bronze/brazing		
Saffire Fluxobronze K for silicon-bronze	Steel, copper, brass	3/32, 1/8 (mm: 2.4, 3.2)

strong joint with no external build-up
(5.9). Traditional bike frame joints
are an example of the technique's use,
though today's aluminum and alloy
steel frames are more often TIG-welded.
Plumber's Yorkshire fittings use the
capillary principle with solder.

Table 5.5 shows common types and
sizes of filler rod for fusion and bronze
work. Each manufacturer's range differs
slightly, but all have the main types
represented. Fusion welding needs
a filler rod of similar material to the
parent plates. If in doubt over which rod
to use, ask a supplier or the rod maker.

Fusion in Practice

For mild steel sheet work up to
0.0591" (1.5mm), start with a number
1 nozzle and ¹⁄₁₆" (1.6mm) mild steel
filler rod. Cut a chunk of plate roughly
6"x3⅞" (150mm x100mm), resting it on
two lengths of angle iron so it's off the
bench and away from any chilling effect.

Comfort is essential, so lean on any
support or (ideally) sit down. Take
the time to unwind any twists in the
supply hoses as fighting these is a mug's
game. Also rotate the blowpipe mixing
chamber so the nozzle points down
at the work. The blowpipe's acetylene
needle valve should be under your
thumb, ready for quick shut-off in case
of backfire (5.10).

Check your goggles. These should be
green-tinted and marked "GW4." Use the
darker GW5 tint if glare is troublesome.

Set a neutral flame. Position the nozzle
so the hottest part of the flame — which
lies around ⅛" (3mm) over the inner
cone's tip — plays on the plate (5.11). If
the nozzle is big enough, within a few
seconds a clear liquid pool will magically
appear under the flame cone. Flying
sparks mean oxides are about. If the pool
is part-covered by a shifting, sparkling
skin, check the flame is still neutral and
that the parent steel is not rusted.

When a pool is reluctant to appear or
flame heat very rapidly burns through
the plate, experiment with flame and
nozzle sizes as follows.

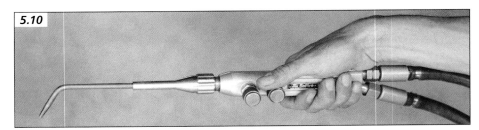

5.10. Turn the mixing chamber/nozzle so you're holding the blowpipe with one
thumb over the acetylene needle valve. This way, fuel gas can be quickly shut off in
an emergency.

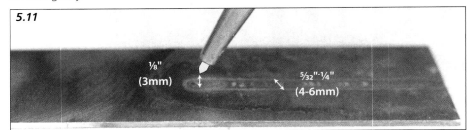

5.11. For maximum heating keep the flame's inner cone about ⅛" (3mm) above the
target. Holding the blowpipe parallel to the work gives the proper nozzle-to-plate angle,
thanks to the nozzle's cranked end.

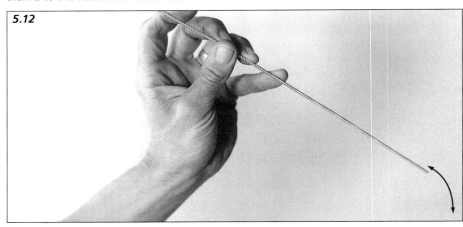

5.12. Filler rod is best balanced between stretched fingers. Feeding in and rocking the
wrist then takes the tip controllably to and from the weld pool (arrow).

Any given nozzle can deliver a range of
heat. To increase it, add more acetylene
via the needle valve and rebalance to
neutral. The flame is now harder —
longer and hotter, so will produce a
molten weld pool more quickly. If burn-
through is your problem, produce a
cooler, softer flame by taking out a little
acetylene and rebalancing.

Understand, though, that there are
limits. The more acetylene you feed
in, the higher is its velocity. Eventually
it's rushing out so fast that the flame
will separate from the nozzle, and the

weld pool will probably be pushed out
of shape by high-speed gas. If either
thing happens, switch to the next-bigger
nozzle: this will give the required heat at
a lower flow rate so the flame and pool
stay happy.

Coaxing a smaller, cooler flame from
a given nozzle means closing needle
valves, thus dropping gas velocities.
Again, watch out: below a critical flow
rate, backfiring will begin. No only
will this blow the weld pool to the four
winds, but backfires are a potentially
lethal precursor to a flashback. *So don't*

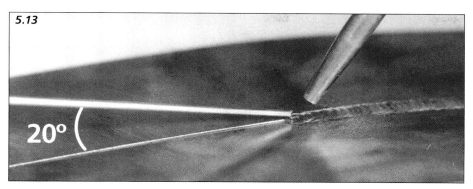

5.13. *Filler rod is held at a much shallower angle than the nozzle, letting it dip cleanly under the latter's tip; around 20° is fine. Keeping filler in the flame envelope during work makes sure it's close to melting point and away from oxidizing air.*

5.14. *A small bird flew over and dropped this. The bead constantly changes shape, suggesting uneven rod/flame work. Fusion was poor, as rod was melted by flame heat rather than being added to a molten pool. Where rod solidified on the underlying plate, cold laps are evident (a). Rather than one continuous bead, unsteady progress gave unfused cold shuts (b). Slapdash preparation left the plate edges misaligned (c), giving little chances of equal fusion.*

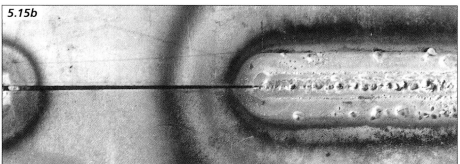

5.15. *Steady work with flame and rod leaves an even-width bead with fairly consistent ripples (a). Edge fusion is good, suggesting that flame heat was about right for plate thickness and speed of travel. The reverse of the joint (b) shows pretty consistent penetration, suggesting careful preparation (plate edges aligned and tacked, in this case with a small gap) and steady, controlled progress. See how heat blueing is equal on both plates, confirming good flame positioning?*

persist when a nozzle starts to pop and bang. If you need a cooler flame, go down one nozzle size.

The aim is to weld with a flame that you are controlling, rather than one that's controlling you. So having found a nozzle/flame size that produces a molten pool fairly quickly, experiment by "pushing" that pool across the plate. Move the nozzle so the pool travels forward, setting a speed that keeps it 5⁄32"-¼" (4-6mm) wide (5.11). Right-handers should work from right to left, keeping the nozzle pointing straight down the line of advance and held at 60°-70° to the horizontal. Left-handers should work from left to right.

When you're happy with this, try adding filler rod. Snip a fresh 1⁄16" (1.6mm) mild steel rod in two and bend one end over so it won't jab in anyone's eye. The bent end is also a clue when you pick it up: the straight end is the hot bit. Righthanders, balance the rod between spread fingers and thumb of the left hand (5.12). Position the hand so the rod lies along the line of the weld and makes a 20°-30° angle with the plate (5.13).

Make a molten pool, and then rock the left wrist so the rod tip dips in and out of the pool. Move the blowpipe and rod together across the plate, dipping as you go: travel to maintain pool width at about ¼" (6mm). Keep the rod low and within the flame's protective outer envelope, sliding it in under the flame's inner cone.

Keep the rod rocking in and out of the pool like a nodding donkey. Try to hit the middle of the pool each time; if you catch the front edge or leave the rod in too long, it'll stick. Don't worry — just hold it still until flame heat frees it.

Whatever you do, don't use the flame to melt and drop bits of rod on to the plates. This only produces the renowned "bird dropping" effect (5.14), which has no fusion and thus zero strength.

Good gas welding is a relatively fast, continuous process in which rod is always added to the center of a molten pool. But only weld with a flame that you control rather than one that's controlling you. The result of tidy rod/flame work looks like 5.15. If yours is closer to 5.16, not to worry. Use the captions to sort out the reason(s). You may need several tries to get the moving pool, dipping rod act together; but once the penny drops it's not forgotten.

Simple Joints

Filler rod isn't always needed. Where design and metal thickness allow, edges can be melted together in a pool and carried along by flame heat. These are autogenous or self-making joints.

Open corner joints in sheet are great for this. To practice, take two 6" x 1½" (150mm x 40mm) bits of ¹⁄₁₆" or 0.08" (1.6mm or 2mm) plate. Tack them into a right-angled roof shape

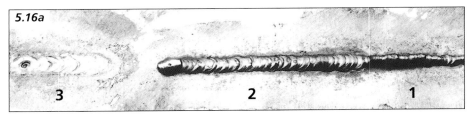

5.16. Travelling too fast or using too small a nozzle leaves a narrow weld pool with little heat available for fusion (a, 1). The resulting bead is thin, humped and not fused into the plate at its edges. Slowing down (or increasing flame size) lets the weld pool grow to around ¼" (6mm) wide, improving fusion and bead appearance (a, 2). But too much heat or too slow forward progress results in a wide, sagging weld area, excessive penetration and a fighting chance of burn-through (a, 3). The reverse of the job (b) shows corresponding penetration: see how it increases as speed of travel comes down. This example was produced by decreasing speed of travel between 1 and 3.

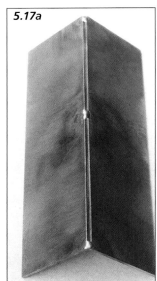

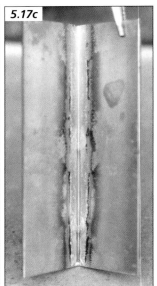

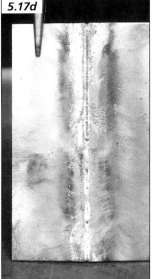

5.17. Open-corner joints are quick and easy in thin sheet. Start by tacking so inner edges just touch all along their length and plates are at 90° (a). Overlapping the edges hides one from flame heat so it can't fuse. Make a molten pool from an end tack and run it down the

joint (b). No filler will be needed if set-up was good. Penetration should be complete (c) and the result can be flattened with a hammer without breaking (d).

(5.17), keeping edges parallel and close together. This is vital, for misalignment will limit fusion and big gaps just let the pool sag though. Make tacks with a neutral flame, fusing filler metal first into one plate then bridging across to the other. Finish the tack by heating the whole tack so it fuses flat into both plates. This is vital or the tack will be weak.

Bring a neutral flame down on an end tack and remelt it into a pool. Now move forward at a speed that allows the plate edges to melt inwards. If you've got it right, a beautifully smooth, rounded weld will result (5.17b). Turning over should reveal even penetration (5.17c).

Butt joints generally need filler rod. Take two more bits of ¹⁄₁₆" or 0.08" (1.6mm or 2mm) mild steel plate, laying them out with a penetration gap equal to (or just greater than) plate thickness. Tack one end as above then regap the plates, which will have moved through distortion. Tack the center and the other end.

Now weld the joint from right to left, using nozzle and flame sizes that let you work quickly but controllably. The plates will want to close up just in front of the flame, which is why a good-sized penetration gap and plenty of tacks are is needed. If you have to wait around for a pool to form or if it wants to freeze on adding rod, there's not enough heat going into the work for good penetration. But if holes keep appearing, you're either working too slowly or with a flame that's too hot. Watch the molten pool very closely: it'll sag just before burn-through, so good penetration is certain if the pool is kept on the point of sagging as it's moved along the joint.

As the joint end approaches you'll have to work faster. Heat has travelled up the

5.18. Building up the finish of a gas weld (arrow) is just as important as with arc. Leaving it hollow is an invitation for cracks to start. Just hold the flame steady and add rod until the crater fills to bead height.

plates, so the finish area will be very hot by the time you get there. Add to this the fact that there's decreasingly little metal left to absorb that heat, and melt-through is a distinct possibility...

so be prepared for it. Speed up as the joint end approaches. At the finish, add filler rod really quickly to build up a pad (5.18). Lift the nozzle straight up when you've done so the weld cools in the

Distortion Control . . . Fat Chance?

All the following applies as much to MMA, MIG and TIG as to gas welding.

Why do welded parts distort? The weld metal itself starts off fluid and in its most-expanded state. As it cools it contracts, pulling at the cooler metal around it. If joint parts can move, they will pull out of line or distort.

Where you reckon that distortion will make trouble, work as fast as you can consistent with good fusion and penetration. If you can get away with it, don't weld all the joint: think about where strength will be needed and share out welds accordingly. If you must weld the whole thing, tack every 2" (50mm) or so. Still no good? Try a lower heat technique like bronze welding or brazing.

A couple of operational techniques help spread heat around. **Skip welding** does just this. Tack up, then chalk-mark the joint into sections 1¼"-2" (30mm-50mm) long. Picking and choosing at random, weld one section then another. The idea is not let heat build up in any area of the joint.

Or there's **backstepping**. Mark out the joint as before, but this time weld the far left-hand section first. Then move to the right, welding one section at a time. Localized heat build-up is not so great as it would be with one continuous right-left weld. Neither of these methods will be 100% successful, but they do help.

Clamping parts so they simply can't move works pretty well. But there is a snag — isn't there always? The bending forces don't go away: they stay locked up in the metal. This usually yields (or stretches) to accommodate them but where it can't, it cracks. Such hidden forces pre-stress the job, using up some of the metal's ability to stand load and maybe making it fail earlier in service than it otherwise would.

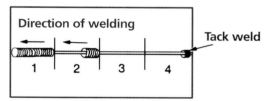

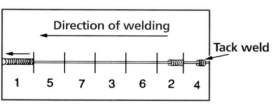

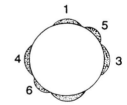

Backstepping (top) helps minimize distortion in any form of welding. Chalk off the sections then weld each in turn, travelling from right to left. In skip welding, sections 1-7 are welded in random order, each again from right to left. When building up a worn shaft or joining a broken one, lay beads on opposite sides to balance heat input and hence distortion (bottom).

One solution is to bring a repair or fabrication up to dull red heat, hold it there for as long as possible then cool it very slowly under sand or a heat-resisting blanket. This should release the stress and ready the part for an arduous life. But be warned: stress relieving is an art in itself. If you think it's essential, ask a specialist. Don't try DIY on any safety-related part.

Distortion builds up as work progresses. When working with sheet, stop now and again and dress down buckled or separated plates with a hammer. Trying to weld these is a fool's game.

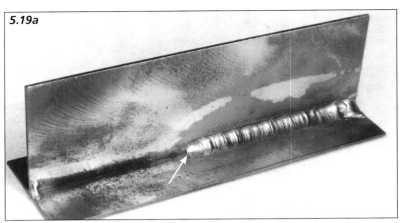

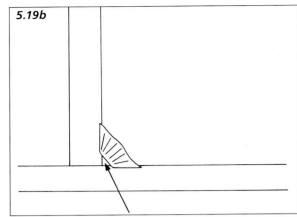

5.19. It's very easy to leave a fillet root unwelded (a and b), especially when the job is resting on a cold bench. But good fusion in the joint walls leaves this T-joint (a) with adequate strength, despite the problem arrowed. Weld one and flatten it with a hammer to see, but be aware that vibration fatigue will destroy an incomplete joint faster than one with 100% root fusion.

flame envelope, which protects it from oxygen and hydrogen in the air.

Flip the finished job over and check penetration. There should be an even bead all along the back. If there's not, either the penetration gap closed right up or your heat/technique were wrong. And you'll find that the plates have bowed — distortion is a big problem with gas fusion welding, as relatively slow travel brings great heat input. Fast work minimizes the problem but can't beat it.

How about a **fillet joint**? The trick here is to supply enough heat, as the joint zone presents a bigger sink than a butt's simple edges. Move up at least one nozzle size over the equivalent butt, adjusting flame size so a molten pool appears fairly quickly and is easily maintained. Ideally, support the joint so it's not resting on anything cold which will add to the heat sink effect.

Tack plates together on side of the vertical plate that's not to be welded, biasing a neutral flame towards the heat-hungry bottom plate. Start the joint itself by building a fillet of weld metal across the plates, making sure it fuses with the parent metal and bites down into the joint root. Then using this fillet as a starting point, move a molten pool slowly along the joint. The vertical plate is most at risk from burn-through and undercut, so be sparing with the flame

up there. Adding filler rod to the top of the pool also helps cool the vertical plate locally.

The vital thing is not to rush. Be certain that fusion reaches right down into the joint root before moving forward. You'll see the pool go hollow and molten metal flow into the root area as fusion happens. If the pool is pushed along before this is seen, all you're doing is building a bridge between the plates: the root won't be filled at all (5.19). This may not matter with thin sheet so long as fusion is good — the bridge is enough. But a joint without root fusion always fails before one with it.

Common fillet problems come from several sources (5.20).

Using too small a flame gives a tiny, fast-freezing pool that won't flow, with the result that root fusion is near-impossible.

Too little heat also gives **cold laps**, where molten metal is just laid on top of the parent plate(s) rather than fusing in. **Cold shuts** are the same problem, only this time happening where weld metal joins weld metal. If the flame is too small or the operator too keen to get on, you'll find trouble where work restarts, on overlapping runs, or where tacks aren't fused into the main weld.

Too much fillet joint heat means burn-through or undercut of the vertical plate. One or both turns up when travel is too slow, if the flame spends too much time away from the joint root, or if a big heat-sink under the work forces the operator to use an oversize nozzle.

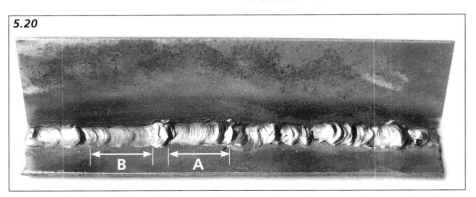

5.20. Not the best of fillets. Using a small flame and travelling at varying speed produced a jumble of poor fusion and misses. But things improved a bit as work progressed leftwards along the joint; short sections (a and b) are better, though undercut is on show at the top edge of (a).

Lap joints are just miniature fillets. Work fast, overcoming the bottom plate's heat-sink effect by using a stronger flame or bigger nozzle than for the equivalent butt. Judge filler quantity and speed of travel so that the upper plate's top edge is just fused in to the joint.

GOING VERTICAL WITH GAS

This isn't too hard. Travel is upward for preference as gravity makes of a mess of the pool coming down, and unlike with arc processes there are no magnetic forces to help hold weld metal in place and direct it to the pool.

Butt joints are prepared as for flat work, leaving a penetration gap equal to or just bigger than plate thickness. Rod and flame angles stay much as they were, only shifted through 90° (5.21). Bend the rod into a hook shape and hold it alongside the joint so the tip points down to the weld pool: this stops burnt fingers.

Use the bottom tack as a starting point. Make a pool from this and add a little rod. Once an onion-shaped hole appears

over the pool you know penetration is complete, though in thin sheet the "onion" might not be necessary. If one doesn't magically pop up, either steepen the flame or physically push it down into the joint root to open up a hole.

When the "onion" appears, move upwards at a rate that maintains it, adding filler rod to the pool as you go. This may sound difficult, but it's easier to do than to read about.

From Soft Solder to Hard

Bronze welding, brazing, silver soldering and soldering are all related. The key difference from fusion work is that the parent metals are never melted. Instead, the filler acts like a glue to bond surfaces to each other. The various filler alloys are arranged on a temperature and hardness scale. At the bottom is soft and bendable solder in its various forms (tinman's, plumber's, pewterer's), containing tin and lead and melting at temperatures from 205°F to 482°F (96°C to 250°C). Silver solders come next, melting between 1,130°F and 1,382°F (610°C and 750°C) depending on what's in them. Joint strength and service temperatures are higher. The bronze alloys (also called "hard solders") are at the top of the temperature tree, melting at around 1,616°F (880°C) — far below steel's melt point.

Overhead gas welding isn't so simple. There's no magnetic transfer of filler, so gravity wants to dump the weld pool into your lap.

This painful tendency is reduced by keeping nozzle size to the minimum consistent with good fusion, and working quickly. Rod and blowpipe angles can be steeper that normal; try bringing them toward a right-angle with the work.

If you're not happy with the notion of welding uphill or overhead, time won't be wasted in dismantling something so you can weld it on the flat. Better spend a while with the spanners than have a job fail in use.

Bronze Welding

A soft, quiet and thoroughly enjoyable pastime is this. Rather than melting parent metals, bronze welding uses a copper/zinc alloy as a "glue" to hold things together or build up worn surfaces. The bond is formed partly from the filler metal hooking into tiny surface features on the joint faces, and partly from controlled local diffusion — but never melting — into the base metal.

Bronze alloys melt at a lower temperature than steel (1,562°F-1,652°F [850°C-900°C] against 2,732°F [1,500°C]) so are good for joining very thin sheet. There are other bonuses: lower heat means less distortion, thick things can be joined to thin things

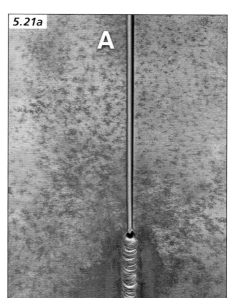

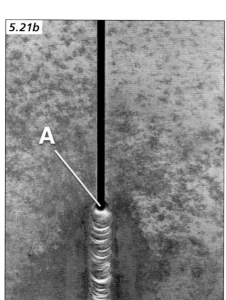

5.21a. Welding vertically up with gas isn't so hard. Use the same rod/flame angles as on the flat. To avoid burning fingers, either hook the filler rod and hold it from the joint side, or hold it above the flame (A). Where necessary leave a penetration gap. 5.21b. To be really sure of welding full-depth, use flame heat and/or physical pressure to blow an "onion" or "keyhole" through the plates (a). For certain penetration maintain the onion as the weld marches upwards.

(5.24) and different metals can be welded to each other.

Five inputs are needed for a good result:

1. The right preparation
2. The right flame
3. The right heat
4. The right rod, and (wait for it)
5. The right flux

Preparation is the single most important thing in any form of soldering, hard or soft. Surfaces must be really physically and chemically clean if the filler material is to flow over and "hook" into them, so elbow grease is called for. Hit rust, dirt or paint gently with an angle grinder, then follow this up with emery cloth. Don't leave the surfaces too coarse or bronze will wet only the high points. Stop preparing only when the weld area is shiny and squeaky-clean. The final job stands or falls by effort put in at this stage.

Flame setting. Though bronze itself hates oxygen, a neutral or just-oxidizing flame is actually used. Why is this? Surplus oxygen in the flame encourages an oxide skin to form over the molten bronze, keeping its zinc content from volatilizing off. But if too much oxygen is around the bronze won't flow.

Heat. The work will be brought only to red heat, so start with a small nozzle and soft flame to control heat input. If the joint won't come to temperature fairly quickly, increase flame size or go up one nozzle. Set very slightly oxidizing flame.

Filler rod type varies with the metals to be joined. All bronze rods are copper/zinc alloys, though exact composition is varied to change their strength and application. **Silicon bronze** fillers are good for general-purpose work on mild steel, copper and brass, although as the latter melts at much the same temperature as the bronze filler, it's effectively fusion welded.

Where more toughness and/or wear resistance is called for, switch to more

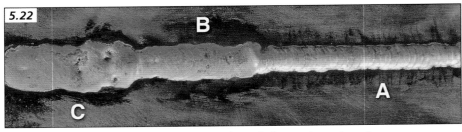

5.22. The right heat input, good fluxing and steady forward travel produces an even-width bronze bead with fine surface ripples (A). Slight overheating pushes the bead out wider and the ripples disappear (B), while overheating drives off zinc in a spitting huff, leaving a pinholed and mottled copper-colored mess (C). Experiment to see how hot bronze follows flame heat over a clean plate. Where the substrate is hot enough and clean enough, there it'll run.

expensive **nickel bronze** rods. The material work-hardens, so is good for building up small worn parts as well as welding steel.

Last of the commonly farm-used rod materials is **manganese bronze**. Priced much as silicon bronze, this offers more tensile strength and is good for joining cast/malleable irons as well as low- and medium-carbon steels.

Flux is essential to chemically clean the surfaces and keep slag, oxides and gases from the weld. No matter whether it's contained in the filler or comes from a pot, the flux had better be fresh. Chuck out any that's gone hard or crumbles away from the rod. A general-purpose bronze flux will do for all three fillers above.

FLOWING THE BRONZE

This is where it starts to be good. Bring those so-clean joint components up to dull red heat. Then start adding bronze. First melt a little on to the surface. Then work along it, dipping the rod to the hot metal to melt more as needed.

If the preparation and temperature control are right, yellow liquid filler will spread quietly and without fuss. Experiment with heat input: see how the bronze only tracks over areas which are hot enough? This temperature-sensitivity lets you keep close control over bronze movement by varying joint heat.

The chances are that first attempts will see the filler spitting and fuming. It's too hot; the zinc is vaporizing in a huff,

5.23. Where work is too cool, bronze sits as a ball and won't flow. The same happens if the plate underneath is dirty.

leaving behind a weak, copper-colored layer (5.22). Less severe overheating doesn't destroy the bead, but the departing zinc leaves weld-weakening pinholes. A carburizing flame produces the same effect.

If a weld shows pinholing, grind off the old deposit and try again. Either use a softer flame, or if this brings on backfiring , switch to a smaller nozzle. If you're already on a number 1, control heat input by playing the flame on and off the work. It might also be an idea to re-check preparation — sharp corners and projections quickly overheat, giving local pinhole problems.

Like other forms of welding, producing good bronze work is like walking a tightrope. Aim for a rapid, controlled process that leaves finely rippled weld surface (5.22, section A). A smooth, featureless bead shows your work was too hot.

Sometimes molten filler just sits around in a ball (5.23). This is telling you that the work is either too cold or too dirty. Either way the bronze won't wet the underlying metal. Re-clean the plates and use a little more heat, then it'll flow where you want.

Dead flux should be cleaned off after welding. The glassy beads of oxide can be chipped gently away (mind your eyes) or, if the work can stand it, be shocked off by cooling in cold water.

Brazing

Brazing uses the same fillers, flame and fluxes as bronze welding. So what's the difference? Bronze welding leaves an external build-up, whereas in brazing all the filler is sucked into the joint by capillary action. So brazing only works where a small gap exists between joint components, for only here can capillary action draw in liquid filler. It follows that the two processes aren't mutually exclusive. If conditions are right for capillarity, a bronze weld can have a brazed region too.

5.24. A bronze welding bonus — you can easily join thick things to thin. This joint's even, fine surface ripples (arrow) suggest temperature was held in check and rod was added steadily.

Excellent preparation is no less vital to brazing. Any dirt will stop the alloy flowing, and if this happens up inside the joint, you'll never know. So prepare the ground by pre-fluxing the components: either sprinkle a layer on them or make flux into a paste and wipe it on.

Then bring both components up to red heat, watching to see that all areas are equally hot. Melt a little filler rod at the joint edge. If the parts are close together, hot enough and clean enough, bronze will be quickly sucked out of sight. Carry on until the joint is full.

The nut and plate example (5.24 and 5.25) shows how thick and thin items can be brazed to leave a neat, zero-buildup finish. Normal surface roughness and fluxing between the bits provides the gap for capillary action.

Cutting, Piercing Holes

Chopping things up is an everyday task. Slicing steel relies on pure oxygen's keenness to combine with preheated metal, giving out extra heat in the process.

The fuel gas can be either propane or acetylene, with propane being the cheaper but more oxygen-hungry option. Here we'll consider only oxyacetylene cutting, as welding requires acetylene fuel gas which is likely to be on the farm already.

All but sheet metal cutting nozzles have a central hole surrounded by a ring of preheaters (5.26). Items are stamped with ANM (Acetylene Nozzle Mix), followed by an imperial measurement ranging from ½" to ³⁄₃₂". This is the central cutting hole diameter, and the bigger the hole, the thicker the plate that can be cut (Table 5.6). Sheet nozzles come in one size only, have a single heater hole and are stamped ASNM (5.27).

The cutting process itself is straight-forward. Let the preheat flame(s) bring the material up to red heat, then use the

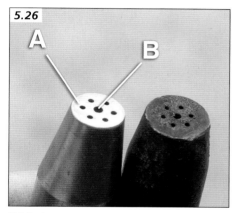

5.26. A new gas cutting nozzle shows off its round, square-edged preheater holes (A) and central cutting orifice (B). With use the edges degrade, but can be returned to pristine condition with a file and nozzle reamers. Keep you nozzles looking like this and cutting will be easy. The same goes for welding nozzle tips.

5.25. Brazing is very strong if done properly — this joint has been attacked with hammer and chisel. The nut was set on a thin bed of flux and brought to red heat, then the flame moved to the plate. With both red hot, a little bronze was melted at each nut flat. It disappeared under the nut to emerge on the thread side. With any bronze work don't spend too long getting everything hot, as oxides will build up and restrict bronze flow.

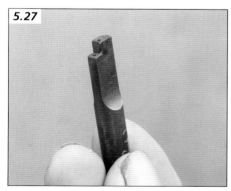

5.27. A sheet metal nozzle has only two holes - the higher for preheating and the lower for the cutting stream. Used with a guide it'll work well in material up to ⅛" (3mm) thick, though cutting heat produces distortion.

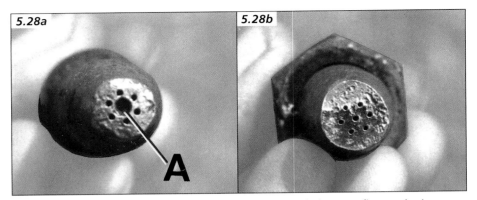

5.28. Bell-mouthing in worn nozzles (a, arrow) causes turbulent gas flow and ruins cutting ability. Nozzle (b) has been dropped, flattening the soft copper alloy tip and deforming holes. It won't cut properly until refaced with a file and reamed.

Table 5.6: Nozzles and pressures for heating, acetylene fuel gas. (TABLE © MUREX)

Nozzle	Approx heat, BTU	Fuel gas pressure		Oxygen pressure	
		Bar	psi	Bar	psi
A-LHT500L	(62kW) 211,550	0.49	7	0.70	10
A-HT 25	(176kW) 600,530	0.35	4	0.35	4
A-HT 50	(309kW) 1,054,350	0.43	6	0.43	6
A-HT 100	(472kW) 1,610,530	0.49	7	0.70	10

Table 5.7: Nozzles and pressures for cutting mild steel, acetylene fuel gas.

Thickness	Nozzle	Fuel gas pressure		Oxygen pressure	
		Bar	psi	Bar	psi
6mm	1/32	0.14	2	2	30
13mm	3/64	0.20	3	3	30
25mm	1/16	0.28	4	3	45
50mm	1/16	0.35	5	3.8	55
Sheet	ASFN	0.40	6	1.7	25

torch cutting lever to feed neat oxygen into the area. The steel actually ignites, turning into liquid iron oxide and giving out more heat to keep the process going. Physical pressure from the oxygen stream clears this liquid away from the cut line, leaving a gap or kerf.

Cutting is the only job in which different oxygen and acetylene pressures are used (Table 5.6). Thus hose check valves must be in good working order!

Compared with welding, gas cutting needs little skill. Instead *a good cut comes down to equipment condition and setting.* Clean cuts just can't be made with poorly maintained gear.

HOUSEKEEPING FOR A GOOD CUT

For quality cutting or welding the nozzle needs to be like 5.26 — holes of the right diameter, all round and sharp-edged. In time the central hole turns bell-mouthed, so the cutting stream isn't blasted out as a straight-edged gas column (5.28). Instead it swirls and tumbles, clearing only a rough, shallow hole. The result is a ragged cut that easily seals up underneath.

Luckily, nozzle makers build extra depth into welding and cutting tips so you can reface them. Do this a little and often, dressing the tip flat with a smooth file and/or fine-grade emery block, then gently reaming the holes. Go steady with

the reamers; you're only aiming to clear away burrs. Don't rasp so hard that the holes end up bigger or turn oval, and keep the reamer parallel so a backfire-producing bell end isn't produced. Resist the temptation (strong though it is) to force in a reamer that's just a gnat's too big, or it'll probably snap off and junk the nozzle.

Proof of a proper refacing job comes on lighting up. Set a neutral flame. With a good nozzle the flame envelope will be symmetrical when looked at from the side and from above. For cutting nozzles, re-check with the cutting lever held down. If the flame is odd-shaped (or if one heater flame shows a feather or is more oxidizing than its mates), first try moving the nozzle round in the torch seat so it seats in a different spot. If this has no effect, grumble and ream again.

With cutting tips, the heat-blue pattern at the face will confirm which hole was misbehaving. See also that a cutting nozzle's internal drillings are clear: with the nozzle held up to the light and angled, you can generally see through them.

Sometimes a smoky acetylene flame appears at the cutting head's nozzle-holding ring. Although not dangerous, this is an irritation which shouldn't continue. The cause is poor seating between the nozzle and head, which can sometimes be cured by very gently dressing the nozzle seats with worn-out emery cloth. But where this doesn't help, there's no option but to buy another nozzle.

If a smoky flame creeps out from the joint between a welding (not cutting) nozzle and the blowpipe, bad seating is again the cause but action needs to be much more immediate. There's a high probability of a backfire, which can jump straight into the nearby blowpipe mixing chamber.

A couple of other points. Keep spare welding and cutting nozzles in a hardwood block, out of harm's way. And because nozzles are made from a soft copper alloy, their internal passages can easily be flattened if you whack them loose from the cutting head. A tap with a softwood block is the kind way.

FLAME SETTING

A balanced flame is the other half of a good cut. Set pressures one at a time with the appropriate blowpipe needle valve(s) open. Table 5.6 shows initial values. If the material to be sliced up is really rusty, increase pressures a little. Outside work on a windy day robs heat, so nozzle choice and setting needs to account for this.

Close all needle valves. Light acetylene first as normal, and adjust flame size so it's "bright." Then open the oxygen valve completely so it plays no further part in proceedings. Feed in oxygen via the cutting attachment's own needle valve until the preheater flames are neutral.

Now hold down the cutting lever. The chances are that the flame balance will shift slightly away from neutral, so rebalance it from the cutting attachment's needle valve. The critical thing is that the flame is neutral when in cutting mode, that is with the lever held down.

A NICE SLICE

To make a cut, hover the nozzle completely over the start point. Keep heater flame cones around ⅛" (3mm) from the surface to maximize heat input and bring the steel to red heat as fast as possible.

Then swing the nozzle clear of the work. Feed in cutting oxygen full-tilt and quickly bring the nozzle back on target. The cut will start cleanly with a splash of molten metal: move off down the cut line. But with thick sections don't move off straight away. Wait until you see molten spatter pouring from under the cut, which shows it's penetrated to full depth. Concentrate hard on keeping the nozzle centered side-to-side over the cut line: any tilt produces a beveled edge which won't usually be wanted. Having said that, it's can help to tilt the nozzle along the line of cut so the tip is

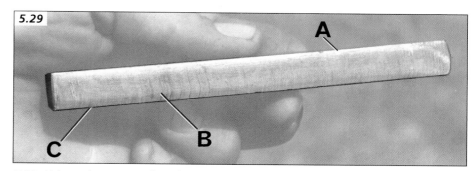

5.29. Halves of a gas-cut plate should separate by gravity rather than needing to be belted apart. A good cut's clean, un-melted top edge (A) shows that preheat flames were not too big or moving too slowly. Draglines just curving at the bottom (B) suggest that the speed of travel was as fast as possible - much faster and the cut would have stopped. A clean lower edge (C) shows heat input was right for material thickness.

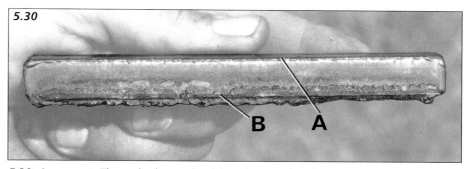

5.30. A poor cut. The melted top edge (A) and crusty slag clinging to the lower edge (B) suggest too much heat, either through using a too-big nozzle, a too-large flame or travelling too slowly. Cuts like this won't part of their own accord as slag holds the kerf together.

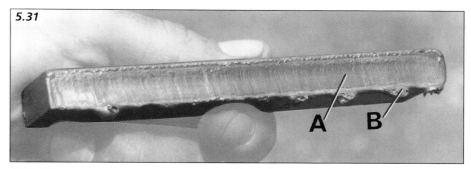

5.31. Where cutting oxygen pressure is too high, the cut face is hollowed by turbulent gas flow (A) and slag again clings to the bottom (B).

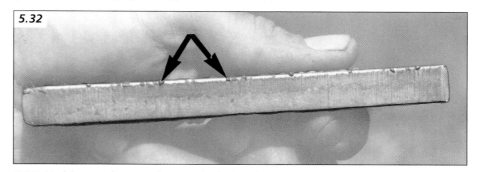

5.32. Nothing much wrong here. Only the beading on the top edge shows preheater flames were too low (arrows).

just forward of the head — experiment and see.

With the cut going well, move as smoothly as the trembling hand allows while keeping those heater cones about ⅛" (3mm) from the surface. How fast do you go? Just fast enough to keep molten metal flooding from the cut's underside.

If travel is too slow the cut's top edges will melt away under preheat. If you go too fast, the cut either stops completely or oxide splashes back out. The first happens when pre-heating is lost, the second when the cut hasn't time to penetrate to full depth. Either slow down, set a longer flame or use a bigger nozzle.

Once a cut has stopped you'll have to pre-heat a little to get it going again. Where the job has wanted and unwanted halves, restart on the scrap side to preserve one clean edge. If you're cutting up to a square corner or want to leave the cut for a moment, make a small loop out into the unwanted material at the change of direction or re-start point. This gives space to rejoin without chewing into the good edge.

Pictures 5.29-5.32 show a good cut and a selection of problems. A clean cut shows smooth drag lines and has little or no oxide on the bottom edges. What there is will chip off easily.

A very common fault is using a nozzle that's too big or a flame that's too long. Both put too much heat into the cut, which seals up with foamy, hard-to-remove oxide. Even the smallest nozzle will cut thicker stuff than you think: try it and see. Most work can be handled by ¹⁄₃₂" and ¹⁄₁₆" sizes if they're in good order.

As with gas welding, a cutting flame can be "softened" by reducing acetylene flow until popping starts, then dialing in a little more again. Softening the flame (then rebalancing it to neutral) cuts the heat input, so is a good ruse for curing underplate oxide build-up in thinner-section plate.

Ten Tips for Faster, Easier Gas Cutting

• Before cutting heavily rusted or painted material, use flame heat and a wire brush to clean it. This makes marking-out much simpler.

• A soapstone stick (5.33) leaves a marking-out line that doesn't vanish under flame heat as chalk does.

• A row of center-punched dots make an easily seen guide for cutting circles. Start the cut near the center, then move outwards in a circle. Slice away until the inner half of each dot disappears, and the hole will be about the right size (5.34).

• Cuts are straighter when a length of angle iron is clamped alongside the line. Choose a size that puts the heater flame cones the right height above the work when you rest the nozzle clamp nut on the upturned edge. Be prepared to make new guides fairly often as heat warps them.

• Start cutting a hole from the center. A punch mark's raised edges come quickly up to heat so target these with the heaters. When edges turn red-hot, lift the nozzle well clear of the plate before gently feeding in the cutting stream — this way the initial splash is less likely to part-block it. Angling the nozzle once the hole starts also lessens the chances of blocking.

• Working in a confined space can quickly overheat a cutting nozzle, causing backfiring. Plunge-cool it in a bucket of warm water, but leave oxygen flowing to stop passages filling up. Check the nozzle ring afterwards in case contraction has loosened it.

• If at any time the nozzle starts to whistle, either spatter or too-low gas flow has led to one or more preheater flames striking back and burning up inside it. Shut off quickly and find the cause, or the nozzle will soon melt.

• Stainless steel or high-chromium welds won't cut, only melt. The stream of cutting oxygen can't combine with them. After all their whole point is to be stainless — that is, non-oxidizing.

• Cast iron will cut, but poorly. Use a carburizing flame and be prepared for a hot, messy job. Let the cut penetrate full-depth before starting, weave to leave a wide kerf and keep the nozzle high to minimize blocks.

• Gas cuts stop very quickly if there's the smallest gap between parts. This works to advantage when (say) slicing a bearing off its shaft or housing, or the head from a bolt while its still screwed home. The downside is that stacks of plates can't be cut.

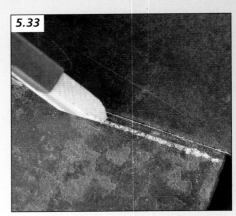

5.33

5.33. A soapstone stick and holder cost pennies. Unlike chalk, the line it leaves doesn't disappear readily under flame heat.

5.34. When cutting circles, punch-dot the center for an easier start. More dots round the edge are easy to see during work. For higher accuracy, make or buy a pointed pivot which can clip to the cutting head. Sit this in the center dot, then use the torch like a pair of compasses.

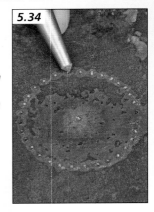

5.34

Section 6

TIG Welding

TIG is the slowest and most skill-requiring of the welding processes. To do it well calls for concentration, patience and fine manual control. Men, with their rush and great fumbling fingers, are outclassed by women on all three counts. So the next time the other half starts looking for a new challenge, shepherd her gently toward the workshop!

Anyway, the TIG sets that are slowly turning up in agriculture have their roots in the 1940s. Then the pressure was on to weld relatively exotic metals—aluminum, stainless steel, magnesium—by a high-quality process, and TIG (tungsten inert gas) did the business. Today's farm requirements are more modest, most likely being steel, stainless steel and aluminum.

Over time, electronics have brought TIG sets down in size, increased their flexibility and dropped the price to something relatively affordable. Several makers now offer multi-role units based on inverter technology, which by putting stick welding and TIG capability into a lightweight and sometimes generator-friendly package, have definite attractions.

If you're not familiar with the process, the panel "What's TIG About?" has the basics. Before getting all excitable it's worth checking both sides of the coin, so for the sensibly skeptical the "For and Against" box puts TIG's arguments in a farming context. Be clear on one thing: TIG is not a replacement for MMA (stick) welding or MIG. Rather, it's a technique akin to gas welding which, if you have the need, can sit alongside existing methods to broaden a farm's repair/fabrication capability.

At buying time the choice is between a DC set and a more expensive and capable AC-DC unit. As many farms are likely to first explore TIG as an add-on to a portable MMA (stick welding) set, we'll deal mainly with the setup and use of single-phase DC units. As there's a lot of kit to sort out before welding starts, we'll begin there.

Setting Up Shop

Kindly nip across and read the panel "Stay Safe." This is important as there's more to TIG safety than meets the eye.

OK? Now to setting up the plant. The tungsten electrode puts the "T" in TIG so comes first, bringing with it several words not usually uttered in the farm workshop.

Electrode type. Originally, electrodes were made from pure tungsten. Though these still have a use, better arc striking, arc stability and electrode durability all come from adding a pinch of thorium, lanthium, cerium or zirconium. For DC welding mild and stainless steels, use a 2% thoriated or 1% lanthanated electrode. For AC welding or for DC welding aluminum, use a 1% ziconiated electrode. Price and quality vary and a poor tungsten will not hold its shape for long, so stick with big-name brands for best effect. Lanthanated electrodes like Murex's Arcrite can be used with both AC and DC: these and ceriated electrodes contain no radioactive thorium.

Identification and storage. Electrodes carry a color code. The standard is green for pure tungsten, red for 2% thoriated, white for zirconiated, gray for ceriated and black for lanthanated (6.1). Whatever their makeup, TIG electrodes are best stored in a box or wallet away from grease and dirt.

Polarity. In DC welding, two-thirds of the heat is generated at the circuit's anode or positive side. To let the electrode run relatively cool and to put most heat into the work for best fusion/penetration, plug the torch into the set's negative outlet for welding mild and stainless steels. But to DC-weld aluminum, use the positive outlet (electrode positive). Much less heat then goes into the work so control is easier and as a spin-off, oxides are part-cleared from the weld. The bad news is that the electrode now has to take two-thirds of the heat, so maximum welding current must be limited and a much fatter tungsten has to be used.

Diameter. Common electrode diameters are 0.04", ¹⁄₁₆", ³⁄₃₂" and ⅛" (1mm, 1.6mm, 2.4mm and 3.2mm). Smaller and larger ones are on offer but these sizes cover normal work. Use the

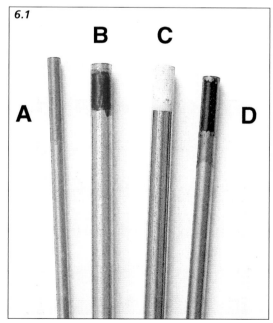

6.1. Electrodes are color-coded. These four are typical — 1% thoriated (A), 2% thoriated (B), zirconiated (C) and Murex's multi-purpose Arcrite (D). Don't grind the colored end or you'll not know what's which.

Stay Safe

Flux is not used with TIG. While that minimizes fume production, particularly high levels of ozone — a lung irritant — are produced by the arc. On top of this, if common platings like nickel, cadmium, zinc, chromium are not ground from the weld zone, very toxic metal fume will result from heating. At minimum, TIG weld only in areas where air interchanges freely. Ideally, take out fume at source by extraction.

Thorium is a low-level radioactive source. Storing and using farm quantities of electrodes is not a problem, but a small risk exists if dust is inhaled when grinding thoriated tungstens. Intact electrodes or those with other included elements pose less risk, so guidelines suggest using alternatives to thoriated items where possible. If not, either buy a purpose-built electrode grinder (see picture) or dedicate a conventional unit to the job. With the latter, either extract dust on-the-go or damp it before sweeping into a sealable plastic container. Dispose of dust in a landfill site.

High frequency (HF)-start sets produce pulses of several thousand volts during arc initiation. Clearly it's daft to put a finger — or any other bit of your anatomy, come to that — between the electrode and work. It follows that an HF-start set is best switched off before changing or adjusting tungstens.

HF voltage can play merry hell with even distant electronic systems. Unplug or remove these where possible before welding on tractors, on implements like a modern sprayer or spreader attached to them, or on dairy equipment. Disconnect a vehicle's battery. Electromagnetic pulses from HF can also

For anyone grinding a lot of tungstens, a dedicated unit like this one from ESAB takes out the guesswork. Dust is caught by a water bath, cancelling the small radioactivity risk from thoriated electrodes.

travel down power cables and are transmitted like radio waves, possibly causing interference with nearby electronic systems.

Last but by no means least, TIG is an arc process. So eye protection demands a quality arc filter — gas-welding goggles offer no protection whatever. Currents are generally low and the arc small, so for good weld area visibility try shade EW9. If you're buying a self-darkening filter, go for one which offers a higher-sensitivity TIG setting range. Higher-current work with aluminum puts out more radiation and light, so move up to shade EW10 or EW11.

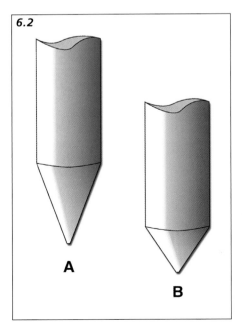

6.2

6.2. Low DC welding currents call for a gentle tip taper (A), while higher amps need a blunter taper (B) — Table 6.1 has details. Grind a tiny flat on the tip to maximize service life.

thinnest electrode possible for a given welding current, as this produces the most concentrated arc and the smallest, most controllable weld pool. New tungstens are usually 6" (150mm) long and can be ground back to about 2" (50mm), at which point and depending on torch design, it gets harder to mount them.

Take Care over Grinding

In AC welding, the heat generated turns the tungsten's tip into a semi-molten ball. Careful preparation isn't needed — just grind a slight bevel on the flat tip.

But in DC TIG, the **shape** and **condition** of the tip are pivotal factors. Although electrodes can be bought pre-ground, you'll need to refinish them sooner or later — and definitely sooner when sorting out your up welding technique.

When will you need to re-grind? After the tip has been contaminated by touching hot weld metal or the filler rod; when a thoriated electrode grows a little nodule behind the tip, or if the arc kicks off at an angle rather than flowing straight down.

6.3. For farm work, a small bench grinder with a fine-grit aluminum oxide wheel is OK for tip grinding. Keep the electrode in line with the wheel's long side, as here. This sets up the tip to deliver the most stable arc — see 6.4.

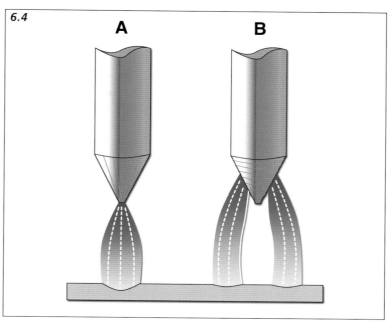

6.4. Grinding lines running parallel with the electrode's length (A) concentrate the arc at the very tip, boosting stability and delivering concentrated heat input to the work. Radial grinding (B) leads to a less stable arc and uneven heat input.

Table 6.1: Current and electrode tip angle

Welding current (A)	Tip angle (°)
20	30
20-100	60-90
100-200	90-120
Over 200	120

Tip shape decides arc stability, and so affects the ease of work and the evenness of heat input. Low currents need a long, thin and pointy tip, whereas higher currents need a blunter taper (6.2). A blunter tip handles more current, gives a narrower arc, keeps its shape better and helps penetration, but starting can be harder and arc wander is more likely.

Use a fine grit aluminum oxide wheel intended for hard metals. As a rough rule of thumb, grind a long 30° taper for currents of 20A or less, and shorten it to around 80° as current goes up. Table 6.1 gives finer detail. Don't leave a needle tip unless arc starting is hard or the job is physically very small. Instead, grind a tiny flat at the end to maximize service life and minimize the chance of

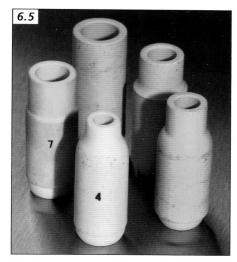

6.5. Ceramic nozzles may carry an identifying number. Match nozzle bore to electrode size for good gas flow.

the point flaking into the weld, where it will form a very brittle inclusion.

And odd though it may seem, exactly how you grind the tip matters. Holding the tungsten parallel to the grinding wheel's long side (6.3) produces mini-gouge marks running towards the tip. These promote a directional, stable arc.

6.6. A gas lens can replace the torch's standard collet body. The fine mesh outlet directs and organizes flow, helping carry shield gas into deep corners or offset draughts.

Holding the tungsten across the stone leaves radial grinding marks which will unsettle the arc (6.4, A and B).

Cups, Cowls or Nozzles— All the Same Thing

Heavier-than-air shield gas is funneled over the weld zone by a ceramic or aluminum oxide cup, cowl or nozzle.

What's TIG About?

TIG (tungsten inert gas) welding brings together elements from conventional stick welding (MMA), MIG and gas techniques. Heat from an arc (A) is used to fuse joint components. This arc forms between the work and a pointed tungsten-based electrode (B), the twist being that the electrode is not melted into the joint. Instead, it remains intact or non-consumed. Where needed, filler metal is added manually from a rod (C).

An inert gas (generally heavier-than-air-argon — [D]) shields the hot electrode and molten weld pool to stop oxidation, and gives a conducting path for current. Gas composition affects arc characteristics and heat.

TIG's small, intense arc suits it to high-quality, precision welding on steel, stainless steel, alloy steels, aluminum and other metals.

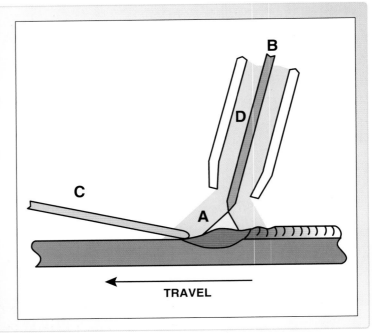

TRAVEL

These come in various shapes and lengths (6.5).

Which nozzle do you use for any given job? If the nozzle is too small for the tungsten and welding current used, it'll glow red hot. Worse, the limited space inside is likely to make the shield gas tumble, which affects arc stability, pulls air into the stream and can expose the weld zone to oxidation. On the other hand, a nozzle that's too big limits sight of the weld pool and doesn't concentrate gas effectively. As a starter, pick an internal bore that's about 4x electrode diameter. Then match nozzle length to the job: shorter ones give better access into tight corners, longer ones carry the shield gas down into deep joints.

Adding a gas lens (6.6) can be a good move where work has to be of top quality or where you're forced to work in a draft. The lens replaces the standard collet body and carries a fine grid through which gas flows, rather than just a ring of holes. Gas reaches the weld in a more orderly way, forming a better blanket round the electrode and pool. Apart from buffering the work against draughts, a gas lens can let you reduce flow (thus saving money) or increase electrode stickout — the

latter boosting joint access and weld pool visibility. The bad news is that a separate lens has to be bought to suit each electrode diameter, and needs a specific gas nozzle to go with it.

Mechanic's Course

So we've got TIG tungstens sharpened, a gas cup selected and a generally expectant air. The next step — putting together the torch — is easier than Meccano, although there are a couple of things to watch.

Picture 6.7 shows how the bits screw together. The sample is an air-cooled torch: water-cooling is generally needed for sustained work with welding currents over 200A, with the main visible difference between units being coolant flow and return lines at the handle.

Check out the nylon nozzle gasket and the back cap's O-ring seal. Both need to be in good order or shield gas will leak and not flow where it's needed. The collet and body/diffuser usually come as a pair, with the required tungsten diameter stamped on the body (6.8). As the back cap is tightened, the collet is pushed into the body and grips the

tungsten. Store collet and body together to avoid time-wasting rummaging.

Back caps come in long or short forms, and it's worth keeping a short one for jobs where access is limited. Screw in the cap until its O-ring seats in the body. Slip the collet inside the collet body/diffuser then run the unit into the torch finger-tight. Slide in the tungsten and add the gas cup, snugging this down on its seal.

Then set visible electrode stick-out. This is the distance between the business end of the gas cup and the tungsten's tip (6.9), which for normal work needs to be ⅛"-¼" (3mm-6mm). Clamp the electrode in place by tightening the back cap. Where you need to see more of what's going on or to reach down into a joint, increase stick-out to a maximum of ⅜" (10mm) — but if you do, think about using a gas lens to be sure that gas shielding doesn't suffer.

Consumables

With filler rod and some **shield gas** you'll be ready to rock 'n roll. Picking the shield gas is a bit of a no-brainer, as pure argon works well with all farming's metals and is the cheapest (though not cheap) choice. Argon is heavier than air,

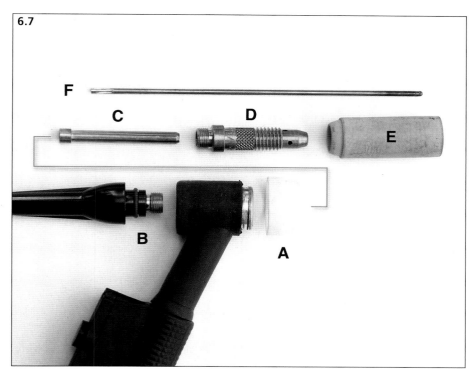

6.7

6.7. The nozzle gasket (A) and back cap O-ring (B) stop shield gas leaking so need to be in good fettle. The collet (C) slips inside the collet body (D), which in turn is finger-tightened into the torch body. Gas cup (E) threads over the body. The electrode (F) is gripped by the split-sided collet as the back cap is screwed down.

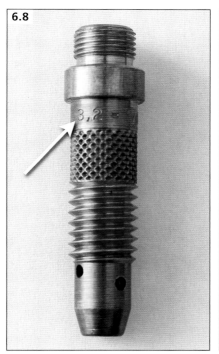

6.8

6.8. Collet bodies have holes for gas exit which must be kept clear. Each is stamped with a size (arrow) matching it to an electrode diameter, and pairs with a collet of the same size.

6.9

3mm-6mm

6.9. Visible electrode stickout should be ⅛"-¼" (3mm-6mm), though it can rise to ⅜" (10mm) for better joint access. If gas shielding suffers at longer stickouts, use a gas lens.

TIG: For and Against

Pro

- Very low current operation possible

- Low distortion

- High-quality welds possible

- Ability to join stainless/aluminum and/or very thin sheet steel

- Joints can be made with or without filler metal

- With filler, can build across gaps

- Slow travel speed helps with complex joint shapes

- Precise control of welding heat

- Small torch can reach tight spaces

- No spatter

Con

- Equipment cost

- Slow compared to MMA or MIG

- Best on material under ¼" (6mm) thick

- Parts must be clean

- Wind-sensitive

- Needs more skill than MIG or stick welding

so in down-hand (flat) work tends to sink helpfully round the weld zone. Sold under various trade names, it comes in the usual cylinder sizes but isn't the only game in town. Mixtures containing a little helium or hydrogen bring better penetration in thick sections, faster welding speeds and lower ozone emission but are even more expensive. For normal work, pure argon is just fine.

Cylinders hold compressed gas at up to 3381 psi (230bar) so the usual safety rules apply. Strap or chain the cylinder to a solid vertical support so it can't topple over. Use a known-good regulator designed for argon and rated for the cylinder pressure, and open the cylinder spindle valve slowly so not to shock the regulator internals. If the unit has a built-in flowmeter, set gas flow at 4-10 liters/min depending on torch gas cup size and any local draughts. If not, set flow to a steady soft hiss at the cup. As with MIG welding, using a flow higher than needed brings no cleaning effect to dirty, rough-prepared work: it just empties the cylinder faster.

What about the **filler metal**? As always, this has to match the material welded.

• **Mild and low alloy steels.** Don't use ordinary gas welding filler rod for TIG, as it won't wet the parent plate very well and the weld pool won't flow easily. General-purpose TIG filler comes in low- or medium-carbon forms, is easier to use, gives a better result and has AWS grade number ER70S-2 stamped on a flattened end (6.10). The more expensive triple-deoxidized fillers are better where the joint can't be completely cleaned, or for joining low alloy steels.

• **Stainless steel.** A wide selection of stainless fillers can be had. For repair and fabrication of common 18/8 or 18/3 stainless steels, go for grades AWS ER308L or ER316L; again the grade will be stamped on the rod. If in doubt check with your welding supplier.

• **Aluminum.** Fillers can be pure aluminum or contain a spicing of silicon or magnesium. Each suits a particular job — pure aluminum for joining pure

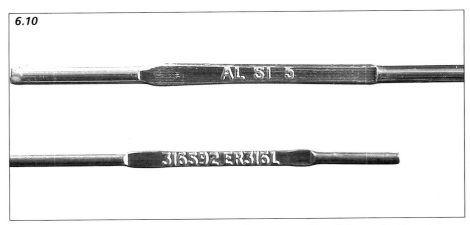

6.10. One or both ends of TIG filler wire usually carries an identifying code, though you might need your specs on to see it. The top rod is aluminum 5% silicon, the bottom one ER316L stainless.

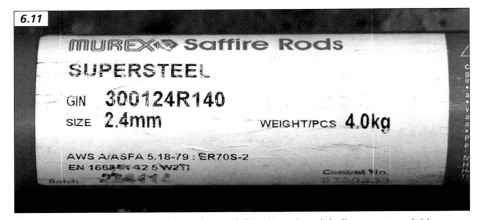

6.11. The rod pack sums up electrode capability. Store it so labeling stays readable, mice allowing.

aluminum (rare in farming), silicon for silicon-containing aluminum alloys and magnesium for magnesium-containing alloys. Unless you're fabricating with new metal it's unlikely that you'll know which aluminum grade you're dealing with, in which case the safest bet is to use a 5% silicon filler like ESAB's Tigrod 18.04 (AWS ER4043). This is the most forgiving choice, suitable for castings, sheet and extrusions. Magnesium-containing filler is more likely to centerline-crack if used on the wrong base material. At a pinch and if the job allows, it's possible to cut a thin strip of parent metal and use that as a filler.

• **Dissimilar and cast steels.** If you need to join mild or low carbon steels to alloy steels or have material prone to cold cracking, try a chromium/nickel filler like ESAB's OK Tigrod 16.53 (AWS grade ER309).

• **Cast iron.** TIG's potentially localized heat input suits it to fusion welding thin-section cast iron. Either find a specific TIG filler or flake the flux from a small-diameter stick welding electrode and use the core wire — but only after very thoroughly cleaning it.

As usual the rod pack is the fount of all wisdom, so hang on to it (6.11).

Take Control

Simple DC sets with either scratch- or lift-arc start won't have complicated controls. Welding current adjustment is usually the only thing.

HF-start sets are more complex, giving the user very fine control over gas flow and current at different stages of the

6.12. Some sets have a forest of adjusting knobs (left), while others back up the buttons with a display for settings and welding current.

weld. This is both good and bad news: good because everything is tunable to suit the work, bad because set-up can get complicated, especially when you're starting out.

Knobs and buttons may be on the set's front panel, or settings might be scrolled through on an LCD screen. Either way, there will be plenty to fiddle with (6.12). Depending on the set, the options could include gas pre-flow, start current, current slope in/up, base current, pulse current, pulse frequency, current slope out/down, crater current, crater time and gas post-flow — *take a deep breath, not done yet* — plus switches to turn HF start on/off, toggle the torch trigger function and allow remote control by foot pedal. AC-DC sets may add wave balance control and more. But here we're only concerned with DC units.

The easiest way to sort it all out is to deal with the peripherals first, then follow the sequence of control events and their effects during a weld. So here goes.

To make torch operation less tiring, some TIG sets let you change the way the torch trigger works. The normal setup is **two-stroke** operation, in which the trigger button is pushed to start welding, held down in work and released to cut the arc or start a slope-down sequence. The alternative is **four-stroke** operation, where the trigger acts as a latch. A push-and-hold starts gas pre-flow and then the arc; then once the arc is running the trigger can be released to make torch control easier. As the end of the joint comes close, a

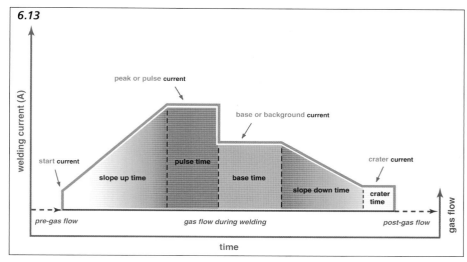

6.13. Time, current and gas flow during a pulsed TIG weld.

second trigger push starts the slope down/crater fill/post-weld gas sequence. Four-stroke triggering is a help once you've got the knack of it.

Better still is a remote foot control (6.14). Plugged into the set and used like a rock guitarist's waa-waa pedal, the remote takes over from the torch trigger and brings stepless variation of welding current during work. Just how it interacts with settings made on the front panel depends on the unit. Apart from giving current variation on-the-move, the remote cuts out trigger operation altogether to reduce fatigue and boost physical control of the torch and arc.

Now for the various adjustments on the set (6.13).

Gas pre-flow sets the time in seconds that gas flows before the arc starts. The idea is to purge the torch and completely blanked the weld zone with

argon, so a pre-flow of 2-3 seconds is about right.

Start current, where adjustable, is set at a level which gives clean, fast arc initiation.

Slope up (also called slope in) sets the time in seconds over which welding current ramps up from start value to the base level. The longer the slope up time, the longer the operator has to see what's happening, position the filler rod and control fusion on a thin edge.

Base or background current forms a team with **peak or pulse current**. Between them they control heat input to the joint, potentially allowing good penetration without burn-through. The idea is simple, though the terminology and systems vary. During welding, pulses of higher current are laid over a lower background current. The extra heat from each pulse produces a deeper

6.14

6.14. A hugely useful add-on, a foot pedal takes over from the torch switch. Use it to start the arc, swell or cut the current during the weld, then trigger the slope-down sequence.

TIG Start: Scratch, Lift or HF?

Starting the arc used to be a potential problem. If the tungsten electrode tip loses its shape during arc initiation then the arc will suffer, and if it sticks to the work the hard, brittle chunk of tungsten left behind upsets weld strength.

The simplest, cheapest technique is **scratch starting**, during which the operator tickles the tungsten over the work and the arc starts as the short-circuit is broken. Tip damage and sticking are both strong possibilities, particularly for beginners.

Slightly more upmarket sets offer **lift arc starting**. This still involves contact between tungsten and the job, but the power supply gives more help. The user either touches and lifts a "live" tungsten, or makes contact then triggers current and gas flow. This causes the set to deliver a precise low-current dose; not enough to spot-weld the electrode to the work, but enough to create a small spark as the torch is lifted. As the tungsten moves upwards the arc begins, and the power supply quickly ramps current up to the full welding value. Lift arc is easier on the operator and brings less risk of tungsten damage.

The most sophisticated option is **high-frequency (HF) starting**. The electrode is poised just above the job and current/gas flow triggered. The power unit delivers a short, sharp jolt of very high-frequency voltage, which generates a shower of sparks between the tungsten and the work. These break down air in the gap, ionizing it so welding current can flow. As there's no physical contact between tungsten and plate the risk of damage disappears, and starting is easier still. HF starting produces a burst of electromagnetic radiation.

melt, during which filler rod is often added. Between high pulses the lower background current lets the pool solidify slightly to control burn-through.

Pulsed current operation is a little trickier to get right than one-step setting, but the added layer of control makes welding easier for the operator. Generally base and pulse current levels are set independently, as frequency of pulsing. To limit penetration in very thin material, set lower, less frequent pulse values on a low background. In thicker work, use higher, longer or more frequent pulses on a higher background. Pulse setting has a big impact on controllability and weld quality, so suck it and see.

Slope out (or slope down) is a decay in current from the welding values, prompted by the operator releasing or re-clicking the torch trigger. Variable from less than one second to several seconds, it's used to limit heat input towards the end of a joint where temperatures are rising fast and the mass of metal able to sink the heat is shrinking.

Crater current and time may be adjustable separately from slope down. Set at a lower value than base current, crater current and duration should be just enough to let the operator maintain and fill the weld finish. If this is not done, the hollow pad left behind will be a focus for cracks.

Post-welding gas flow is also important to resist cracking and to let the tungsten cool in an inert shield. If gas is cut off as the arc is broken, the molten weld pool will take up nitrogen, hydrogen and oxygen from the air and turn brittle. Setting gas post-flow time to 3-4 seconds is enough for the weld pool to solidify in thin material… and naturally the torch has to be held over the weld until flow stops.

Clearly more complex DC TIG sets have plenty to play with. So even more than with MIG, it pays to understand the basics then experiment while you're learning: change one variable at a time to see its effect, and then how different settings interact. And as with any technique, don't even think about welding safety-related items unless you're completely sure of the outcome.

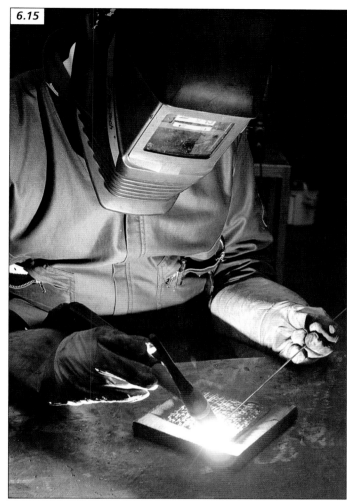

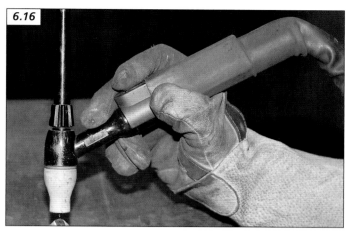

6.16. Grip one: Holding the torch like a pencil with your index finger on the trigger gives most control, though it's harder initially.

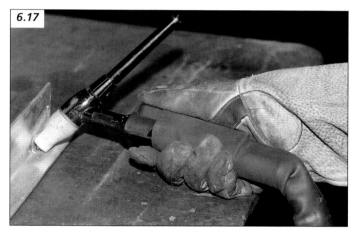

6.17. Grip two: The "handshake" grip with thumb on trigger helps with T-fillet joints.

6.15. Physical relaxation is the first step towards control. Angle the work so it falls comfortably under your hands; rest your forearms or elbows on anything that's handy, loop the torch lead over a support so you're not fighting its weight.

Making the Weld 1: Comfort First

TIG is second cousin to gas welding, in that the operator has to control a heat source with one hand and probably add a filler rod with the other. If you can gas weld, you're off to a flying start with TIG. Mind you, gas is less demanding—the flame cone can touch the work without major consequences, and if the weld area gets too hot you can lift the flame to let it cool a little. TIG isn't so forgiving, so operator skill must rack up another notch.

Self-help is the place to start. All welding is easier if you're physically relaxed and not fighting the equipment, which produces the first rule of TIG: *get ultra-comfortable before you begin*. If you haven't got a welding bench, knock one up and bring portable jobs to it. There you can rest any or all of your anatomy on props assembled for the purpose— a stool for the bum, support for the elbows, something to take the weight of the torch leads (6.15). Angle the job so it falls naturally under your hands, and try relaxing your whole body before starting the arc. Unless you're so relaxed that you fall off the stool, you'll find control is much easier.

When you must to go to the work, stick to the same principles. It may take a while to get settled, but better that than rushing: torch control can be hard enough without external complications.

A key element in comfort and control is the way you hold the torch. There's no right or wrong here, just alternatives. One tack is to grip it like a pen, with the index finger working the on/off button and the neck supported by the fourth and little fingers (6.16). This takes a bit of getting used to, though the payoff is better control and easier change in electrode angle. The other option is to just grab the torch handle like you'd shake somebody's hand, using the thumb for the trigger (6.17). Easier to begin with, this grip is also best for fillet

6.18. TIG gloves — made from thinner leather than normal welder's gauntlets — give better feel.

joints. Whichever way you choose, hold the torch gently (as tension generates wobbles) and invest in a pair of thin-leather TIG gloves for better feel (6.18).

PREPARATION IS ESSENTIAL, NOT OPTIONAL

Workshops steeped in stick welding tend be a bit cavalier over preparation. A good rod flux usually floats gunk from the weld pool and locks it up in the slag, leaving the weld wholesome (or wholesome enough).

TIG fillers have no flux; just different levels of de-oxidants, depending on specification. And the process's argon shield gas only keeps weld-weakening gases in the air at bay. So how much surface contamination can TIG tolerate? Very little, and if you're after a top-quality result, none at all.

MIG's litany of grot-removal remedies applies to TIG. A very little surface oil or grease is OK; it should vaporize in the approaching arc heat. The thicker the layer, the more likely it is to spit and contaminate the tungsten. You'll see and

hear this. A TIG electrode contaminated from any source produces a misshapen arc, which in turn means uneven joint heating, difficult weld pool control and potentially poor fusion and penetration. If you see trouble don't carry on. The tungsten tip will discolor too, perhaps turning blue. A last-minute wipe of the weld area with acetone will shift light residual oil and leave a good surface.

Of course paint must be ground off completely. Where that's not possible, heat the joint area with a gas flame or the TIG arc until the paint blisters, then wire brush the weld zone clean. Not ideal but better than nothing. Platings can contaminate the weld, and some — yellow cadmium and galvanizing's zinc especially — are potentially evil when inhaled; as said before, cadmium fume can be fatal. Grind platings right off and avoid breathing the dust. If you've a lot of work to do with galvanized steel, think about using TIG brazing rather than fusion welding; more on that later.

Rust tolerance depends on how much of it there is. Flaky stuff go, no argument. Filler wire de-oxidants should float very light surface rusting from a joint, though if there's any doubt and where maximum joint strength is essential, either abrade rust away or switch to triple de-oxidized rods. On thin sheet steel an emery flap wheel is kinder than an angle grinder.

When preparation has been skimped or the wire can't cope you'll soon know. As with gas fusion welding, a warning sign is a sparkling scum of oxide around, and sometimes on, the weld pool. In more severe cases the tungsten will pick up contaminants or the weld will turn visibly porous, in which case the only option is to grind it out and start again. Unlike other arc processes, a TIG arc running in good conditions is almost silent and produces no sparks.

KEEP GASSING

Which brings us, if only by the most roundabout of routes, to shield gas flow and working outdoors. The first truth is that as with MIG, winding up gas

flow can't (and won't) make up for poor surface preparation, and is likely to make things worse. Argon acts only as a blanket to keep air off; it can't react with weld pool contaminants and clear them away. So how much flow is enough, and why is too much a bad thing?

Argon regulators are usually calibrated in cubic feet per hour (liters/min) (6.19). Suggested flows run from 8.5 to 25.5 cubic feet per minute (4 to 12 liters/min) or more, depending on material type, plate thickness and joint design. Rough-set the regulator then fine-tune flow with the torch's trigger pressed.

Lower flows go with the smaller weld pools and lower currents used to join thinner material, while bigger pools needed by thicker sections need more gas to shield them. Aluminum needs higher flows than steels, as it's welded with chunky electrodes and their relatively wide weld pools. Flat-surface joints like butts and laps, along with outside corner joints, are more sensitive to low flows than fillets and inside corners, which through their cupped-hands shape tend to hold the gas blanket in place.

Table 6.1 gives general recommendations. Too little gas shows itself in several ways: the arc can wander if air gets alongside it; the weld pool is harder to form and carry

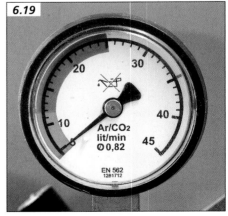

6.19. Argon regulators often show gas flow in liters/min. Adjust flow with the set turned on and the torch trigger pressed, or backpressure will give a false reading.

Table 6.2: Material thickness, tungsten diameter and gas flow

Plate thickness	Electrode dia	Argon flow (cubic feet per hour)
(1mm) 0.035"	(1mm) 0.040"	(4 l/m) 8.47
(2mm) 0.0787"	(1.5-2.0mm) 0.0591-0.08"	(4 l/m) 8.47
(3mm) ⅛"	(2mm) 0.08"	(5 l/m) 10.59
(4mm) ⁵⁄₃₂"	(2mm) 0.08"	(5 l/m) 10.59
(5mm) 0.2"	(2.0-3.0mm) 0.08"-⅛"	(5 l/m) 10.59

Table 6.3: Tungsten diameters and suggested welding currents

Tungsten (mm)	DC		AC
	Electrode -ve	Electrode +ve	Zirconiated
0.020" (0.5mm)	5-20	n/a	5-20
0.040" (1.0mm)	15-80	n/a	15-80
1/16" (1.6mm)	70-150	10-20	70-150
3/32" (2.4mm)	150-250	15-30	110-180
1/8" (3.2mm)	250-400	25-40	150-200

Note: *Values for DC welding apply to thoriated, ceriated or lanthanated electrodes, used with argon shield gas.*

forward, and the weld turns porous from oxidation. Those are extremes; a more common sign of too little shield gas is blackening of the weld, caused by contact with atmospheric oxygen.

At the other extreme, too much flow produces high velocities in the ceramic. Turbulence can then pull air into the argon stream, again producing an unstable arc and porosity. In extreme cases the gas column will physically push the weld pool about, just like a gas flame that's too hard for the nozzle. If that's happening, do the obvious — reduce flow at the regulator or switch to a bigger cup.

TIG's requirement for consistent gas shielding make it very sensitive to drafts, even more so than MIG. If you have to work outdoors, use a gas lens rather than a straight ceramic and put up as much physical cover round the job as possible.

ANGLES, DISTANCES AND SPEEDS

Basic torch and rod angles for DC TIG on steels are in 6.20. A fairly steep torch angle fires arc heat directly into the joint, rather than letting it play ahead of the molten weld pool and risk melting the filler early. Keep the filler rod low and close to the action, so it stays inside the protective gas shield and can be dipped squarely into the front edge of the molten pool. If exposed to the air, the filler wire tip will oxidize and take this contamination into the joint.

Hold arc length relatively short at ⅛"- 0.2" (3mm-5mm), depending on the current. The longer the arc, the wider and cooler the weld pool.

6.20. Angles for down-hand (flat) TIG. At the start of a joint the torch can be steeper than here; moving the tungsten in a tight circle helps fuse joint edges into a molten pool. Once this is established, tilt the torch back a little and set off down the joint line. Filler rod (right) is kept low, with the tip never far from the arc so it stays in the protective shield gas.

CURRENT SPOTTING

Without a foot pedal, the scope to vary heat input while TIG welding is limited. Varying travel speed is one way (as slower=hotter), changing the volume of filler added is another (as more filler=cooler). Both can compromise weld quality if taken too far. The thing you can't do is lift or lower the torch more than a little, as the power source will outwit the operator by adjusting arc current to try to keep heat input constant. The only proper way to vary heat input with TIG — and then only where the set and job location allow — is to add a foot control (6.14), which works as a throttle pedal to start, stop, swell or fade welding current as you work. It can even be used as a crude pulse generator if the set doesn't feature the option.

ELECTRODE DIAMETER

Like the way you hold the torch, there is no "right" diameter for any particular job. Rather, various diameters can be used, as long as the current needed to produce a stable arc and a controllable weld pool lies in the tungsten's working range. There's the physical side, too — fatter electrodes and bigger gas cups can't get to limited-access joints.

Electrodes can tolerate a surprisingly wide current range. But recognizing the maximum and minimum points still matters, as these prompt a change in diameter. With thoriated electrodes on mild and stainless steels, the warning signs are:

• **Current too high for diameter.** The electrode starts to lose its fine point during work, followed by complete physical collapse.

• **Current too low for diameter.** Poor arc ignition, then the arc tends to wander. A weld pool won't form easily as the tungsten absorbs too much heat.

AC/DC welding aluminum needs zirconiated electrodes. For these the signs are:

• **Current too high for diameter.** The electrode's ball end wobbles or comes off.

• **Current too low for diameter. The** arc can't break the weld pool's oxide film as the tungsten is absorbing too much heat.

Making the Joint 2: Bringing It All Together

The simplest joints are those that don't need added filler metal. They're a particularly good place to begin, as both the operator's hands are left free to support the torch. That's always a positive thing to do where you can, as most of us need all the help we can get with any sort of welding. It's also a useful way to develop control — start with two hands, take one away.

So, filler-free or autogenous (self-making) joints are fast, neat and often used in TIG production work. Common examples are the lap joint (6.21) and outside corner (6.22). On farms the first is handy for repair, the second when making things like guards.

As always, laps and outside corners need accurate alignment before welding. The plate edges must touch all the way along the joint line, for the technique relies on melting equal parts of each to form the joint. Where there's a gap, the weld pool wants to disappear and one plate edge tends to burn back (6.23). So start by straightening the plates, then tack them at most every couple of inches and close any gaps with gentle hammer work.

If you're using ¹⁄₁₆"-0.08" (1.5mm-2mm) mild steel sheet, start with a ¹⁄₁₆" (1.6mm) thoriated tungsten and about 50A. On sets that allow it, use the following: 1sec pre-gas time, HF start, 20A start current, 1.5 sec slope up, no pulse, 1.5 sec slope down, 15A crater current and 3 secs post-gas time. With simpler units, dial in as much of that as the set allows.

To run the weld, first find a comfortable working position. If you can reach the whole joint without shifting, so much the better. With the torch vertical, start the arc and close it down so the tungsten is about ⅛" (3mm) from the plate. If the joint halves don't melt equally, circle the torch to encourage them to form a central pool. If that pool doesn't appear quickly — remembering to allow for

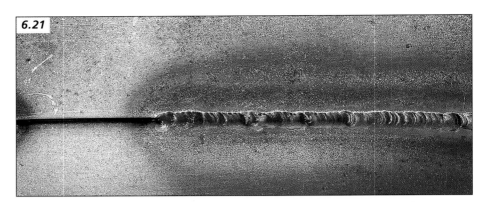

6.21. Lap welds don't necessarily need filler rod, as long as the plates are in close contact with each other. Setup: ¹⁄₁₆" (1.6mm) thoriated tungsten, ¹⁄₁₆" (1.6mm) mild steel sheet, 60A.

6.22a. An outside corner tacked and ready to go . . .

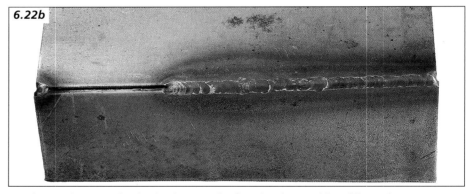

6.22b. . . . then part-finished. Edges can be fused without adding filler rod.

slope-up time — then stop and dial in a little more current until it does.

Once the pool is established, tilt the torch back a bit and move off down the joint. Travel at a speed that lets the pool melt the plate edges (for an open corner joint) or the upper plate's top edge (for a lap joint). Heat will get to the far end of the plates before you do, so take your finger off the torch button to

start the cooler slope-down sequence before the finish. Hold the torch steady over the finish area — the pre-set lower crater current should stop burn-through — then once the arc goes out, keep the torch there until the post-weld gas flow stops. That last bit is important. If air gets to the weld finish before it cools, cracking is likely. The continuing argon flow also keeps air from the tungsten while it too calms down.

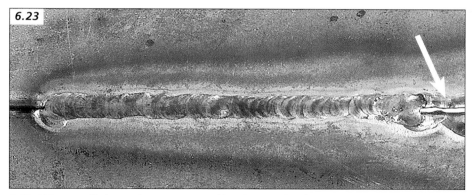

6.23. Whoops. If plate edges aren't touching it's very hard to start a joint without filler (arrow). The weld pool will disintegrate if there's a gap part-way along, too.

NOW ADD SOME FILLER . . .

Next step is to wield a filler rod. Butt joints are used everywhere so that's the job — weld one plate to another laid alongside it, in the flat or down-hand position. As with laps and opens corners, tack the plates together first. This time, though, leave a penetration gap roughly equal to plate thickness (6.24). It won't matter if the edges aren't straight and tidy, as filler will take care of them. How thick should the filler rod be? Chunky enough to build up the joint without non-stop dipping, but not so thick that adding it causes the temperature to dive and the molten weld pool to nearly freeze.

The technique is this. Form a pool, add rod, use arc heat to push the pool forward into fresh metal, add rod, push the pool forward again…and so on. Shift the tungsten back a little (but don't lift it more than a tad) before adding rod, so there's less chance of a filler-tungsten collision. If that happens you'll have to stop and re-grind the electrode tip, as the arc will no longer heat evenly.

Start the arc, make a pool and move off. Try very deliberately to place the filler rod directly into the pool, just like dipping a pen in an inkwell. If you're too young to know about inkwells, think dipping the business end of a chip into a pool of ketchup. Doesn't matter if the filler sticks to the pool's front edge; leave it there and arc heat will quickly free it. The thing not to do is let arc heat melt the filler ahead of the pool. If that

happens, filler metal will just fall on the plate rather than fully fusing with it.

Travel speed depends on the balance between welding current, material thickness and filler addition, but as usual the idea is to work as fast as possible while still keeping control of the molten pool. As the joint end hoves into view, trigger the slope-down sequence.

Make a tidy finish, adding filler so the weld bead isn't left concave and weak. Wait for post-weld gas flow to finish, and job done (6.25).

If your set allows, experiment with the two-stage current available through pulse welding. Try different pulse sizes and durations, using the high current for the melt-and-push-pool-forward phase and the cooler background time for adding filler. Pulse welding adds a lot of flexibility, and is very useful to balance penetration against burn-through on thin sheet.

. . . AND FINALLY FILLETS

Number three common joint is the T-fillet, welded in the flat (down-hand) position. The basics are the same as always: tack up, make a pool from both joint halves, add filler low and to the front of it, trigger slope-down before the end. Several detail tweaks should make this go easier.

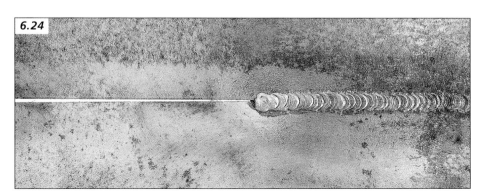

6.24. Butt joints in mild steel sheet need a penetration gap about as wide as the plates are thick. While it's possible to weld these without filler, penetration will be shallow. This one used 1/16" (1.6mm) rod, 60A.

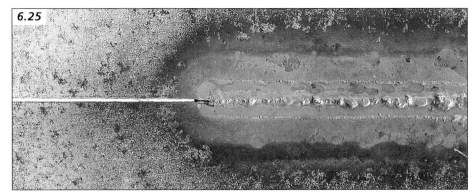

6.25. The reverse side of 6.24 shows that where the penetration gap, heat input and speed of travel are about right, fusion will be full-depth.

As usual, get comfortable and use anything handy to take torch weight. Shift your grip to the handshake style (6.17), which gives better control for fillets. If you're of an economical turn of mind, gas flow can be reduced a little as joint shape maintains blanket coverage better than butts or laps. For best access and visibility, switch to a smaller ceramic and increase tungsten stickout to a maximum of ⅜" (10mm). Pick a filler diameter that looks as if it'll build the required weld fillet without burning back like a fuse, or you'll need shuffle rod through your fingers too fast.

Tack and dress the plates so the joint edges are touching, particularly with thin sheet. Gaps expose the vertical plate, which will burn away quickly. If this looks to be starting, drop the temperature locally by adding filler at the top of the joint.

Build a pool between both plates and move off. Keep the electrode angled into the root of the joint (6.26) rather than pointing it too far forward down the line, as that will skip heat forward rather than putting it where wanted.

Use the standard technique — melt, push forward, add rod — but be certain to hold the melt phase for long enough to let metal right at the root of joint fuse into the pool. Again, pulsed operation can help with thin sheet. If you don't get a proper melt (and it's easy not to) then you're just building a bridge between the plates. This may do for light-duty applications, but will fail earlier than a fillet welded to full depth.

Finally, watch out for undercut. This first shows on the vertical plate as a weakening notch above the weld. It's the outcome of too much heat and too little filler rod. To fix undercut? Drop the welding current, work a little faster or add more filler high in the joint to protect the vulnerable vertical plate.

THE UPS AND DOWNS

To weld vertically upwards with TIG, take the normal flat technique and rotate it through 90º, as you would for gas welding. That puts the torch pointing up the joint line, and the filler rod pointing down it (6.27). As with flat T-fillets, switch to the "handshake" grip

and use a little more electrode stick-out to reach into the bottom of the joint.

You can't build steps to support fresh weld metal as you can with MIG or stick work, as the weld pool is too fluid. So the pool has to be carried up the joint non-stop, which takes good physical control. As with other TIG jobs, pulsed current buys time, especially on thinner metals. Collect yourself in the cooler background pulse, then use the hot pulse to add road and push the pool into fresh parent plate. Keep the filler wire low so it doesn't tangle up with the tungsten, and plonk the tip directly into the pool's leading edge. Which

6.27. Vertical TIG is about as hard as it gets. Rest body weight, relax; be happy. Travelling up or down, use the same filler/torch angles as you would for a flat butt. A relatively thick filler minimizes rod additions. If your set allows it, try a fairly long background current pulse to give a mini-breather between the higher-current pushes into fresh plate. Setup: ¼" (6mm) mild steel plate, ⅛" (3.2mm) filler, ⅛" (3.2mm) thoriated tungsten, 180A, no pulse.

6.26. Basic torch and rod angles for a flat (down-hand) T-fillet. Feed the filler rod in low so it can slip under the arc. Keep the tungsten relatively square to the joint; heat then goes where it's needed and filler — which is delivered straight into the weld pool — isn't melted early.

can be rather easier said than done, particularly where the job doesn't make it easy to rest on something solid.

Current and rod diameter? As ever, heat input must walk the line between a good, fairly fast melt of the parent plates for penetration, and producing a hot, fat weld pool that's easy prey to gravity. If the pool sags and burns though, drop the current. If the pool is hard to start and then maintain, add amps. Pick a fill rod diameter that builds the joint readily without constant dipping, as the more you dip the more control you'll need.

Welding vertically down with TIG can be very handy. Gravity can work for you, pulling down the molten pool so it's constantly quenched by fresh plate. It's often the best technique with thin sheet, generating strong results quickly without the need to use filler, which means that two hands can hold the torch.

MATERIAL CONSIDERATIONS

Now for some specifics on different metals. The **steels** (mild, low carbon and stainless) can all be welded easily without filler rod if the joint allows. Watch out with no-filler butts, as complete penetration is hard to achieve without using a lot of heat. Such welds have a shallow profile and centerline-crack easily, as there's not the thickness in the joint to handle stress from contraction or loading.

Tungsten composition and quality influences ease of work and joint integrity. For steels, a good all-round choice is a big-name brand of 2% thoriated electrode. Be specific when buying and if necessary spend a little more, rather than taking what comes. Multi-purpose and 1% thoriated tungstens tend to lose their tip easier, particularly in scratch-start DC TIG. In all cases — mild and stainless steels — pure argon shield gas is the best choice.

Heat travels slowly through stainless steel, which expands and contracts more than the mild stuff. As a result

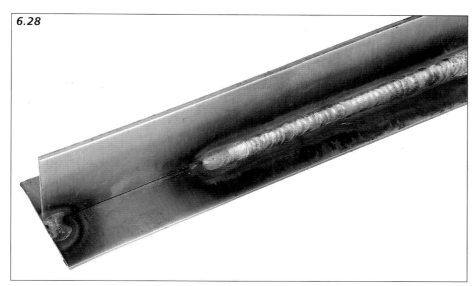

6.28. Stainless steel TIG-welding with good gas shielding and about the right current shows even, small ripples and a straw-color finish. Setup: 0.0591" (1.5mm) stainless sheet, 1/16" (1.6mm) thoriated tungsten, 1/16" (1.6mm) filler, 45A base current, 55A pulse.

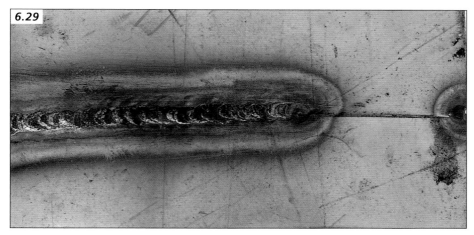

6.29. Underside of a butt joint in stainless steel, welded with too much current. Surface has typical burnt, black oxide finish.

stainless is much more likely to distort, so that needs thinking about before starting a job. Putting in as little heat as possible will help; use pulse welding if you can. In any case, set the minimum current that gives good fusion and allows steady, consistent progress, and keep the filler rod dipping in and out regularly to balance pool heat. Using a gas lens is a good idea, as the wider, more even argon blanket it delivers makes a better job of excluding oxygen which, when it hits hot weld metal, turns the surface rough and black.

When you've got the balance right, the joint will show even ripples and have a fine straw color, at least on the top side (6.28). Underneath it'll probably be burnt-looking from oxygen attack, but not to worry; that's mainly a surface effect that doesn't weaken the joint (6.29). If it bothers you, cleaning with weak acid will get rid of most of it — a good welding supplier will suggest what to use. Or to beat the problem completely, make a purge box — one with holes placed to flow argon to the underside of the weld zone, keeping air at bay.

6.30. Aluminum must be cleaner than clean before welding. A zirconium flap wheel or stainless steel wire brush whisks off oxides without leaving contaminants behind.

The cleaner stainless steel is before welding, the better finish you'll get. It's more contamination-sensitive than mild steel.

We've touched before on DC welding aluminum, suggesting that (with a bit of luck and a following wind) it can be used on non-critical jobs, though never where safety is involved. Here's some explanation and few thoughts.

The problem in using DC TIG to join aluminum lies in persuading two molten pools, one on each joint plate, to fuse into a single pool which can be carried along the weld line. Stopping the happy wedlock is a skin of very hard-to-melt refractory oxides on each pool, with the result that it's a bit like trying to join two footballs full of water. The buzz of alternating current (AC) easily breaks that skin, so AC is the preferred choice. But if all you have is a DC set then the following is worth trying.

DC negative polarity (i.e., the torch plugged into the negative socket, as usual) isn't good at breaking the oxide skin or cleaning oxides from the plate surfaces. Reversing the leads to make the tungsten DC positive is better, though nowhere nearly as effective as AC. The snag is that now the electrode carries two-thirds of the arc energy while the work takes only one third, so a fat zirconiated electrode is needed to stand the heat. This heat can be considerable, as although aluminum melts at a much lower temperature than steel, it conducts heat much faster so relatively high input will be needed.

Really thorough preparation is vital when trying to DC weld aluminum. The cleaner the metal, the easier it'll form a weld pool. Pre-heating up to about 212°F (100ºC) with a gas torch will make surface oxides easier to shift, followed up very quickly with a good scrub from a stainless steel wire brush or stainless wool pad. Using mild steel products will likely add iron oxides, so will hinder rather than help. More violent cleaning comes from an aluminum oxide flap wheel or disc (6.30). Don't use a disc designed for steel: it'll only leave weakening inclusions. Keep filler rods clean and don't wear dirty gloves, or the filler will pick up contamination as you handle it.

Grind a zirconiated tungsten square — it'll develop a rounded tip in use. Hold the torch almost vertical, tilting it just enough to see the front edge of the weld pool. The arc from a blunt tip isn't as directional as the arc from a pointed one, and too much tilt elongates the pool and makes breaking the oxide skin even harder. If the torch angle is too shallow the filler rod will also melt too soon as it approaches the weld pool.

To help break the skin and get the pools to fuse, work the arc round in a tight circle as you start the weld. Once the pools have joined, the surface will clear to reveal bright, liquid aluminum and you're away.

Pure argon shield gas is fine. If you're only going to keep one sort of filler, go for a 10% silicon type like ESAB's OK Tigrod 18.05 or Murex's Saffire aluminum 10% silicon. These match most common sheet and castings and are usefully ductile, so are less prone to cracking as the weld solidifies.

TIG BRAZING — OFTEN A SAVIOR

When technique is less than perfect (and whose isn't?) joining thin steel sheet can give anyone a hard time, even with TIG's low currents and tiny arcs. That's when TIG brazing can lend a hand. It's also good on galvanized steel, where the heat of fusion welding destroys the coating, puts poisonous zinc fumes into the air and generally isn't much cop unless all the plating is ground off first. TIG brazing can run over galvanizing without much upsetting it if heat input is minimized, making for faster, more pleasant work — and the joint will need just a coat of zinc-rich paint to protect against corrosion.

What's more, TIG brazing's relatively low temperatures deliver low distortion welds in mild steel. And it can join cast iron, silicon bronzes, brass and dissimilar metals, so is a very handy technique to know.

The idea is the same as gas flame brazing or bronze welding. That is, an alloy with a lower melting point than the parent metal is used as a metal-based "glue." Brazing uses capillary action to suck the molten filler into a joint, so the TIG version isn't strictly that. Rather it's braze welding, where the filler metal is flowed over hot, clean surfaces to build a fillet of material (6.31).

The preliminaries parallel gas brazing. The parent metal(s) must be completely clean or the filler will blandly refuse to flow, so the usual thorough degreasing/light grinding routines apply. And the joint metal has to be hot enough, or filler laid on its surface will pull into a tight ball and go nowhere.

The filler rod can be silicon, aluminum or manganese bronze, but MUST be specified for TIG work — gas filler definitely won't do. Silicon bronze is the most ductile and flexible in service. The weld it leaves is more copper-colored than aluminum bronze's yellow finish. Manganese bronze gas rods can be used with TIG to join cast iron, if the joint faces are buttered with a layer of filler before the weld is made.

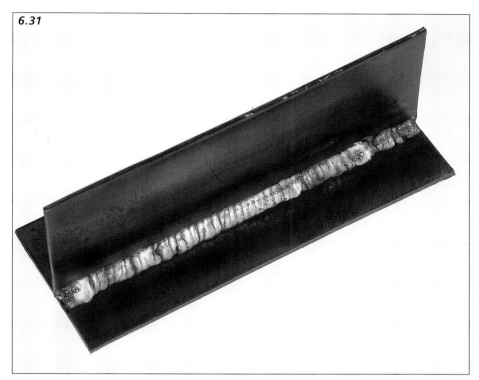

6.31. TIG brazing is good backup to fusion TIG for joints in very thin mild and stainless sheet, and in brass and bronze alloys. It's also good on galvanized steel, reducing fuming and coating loss. Yellow finish is typical. Setup: ¹⁄₁₆" (1.6mm) mild steel sheet, ¹⁄₁₆" (1.6mm) thoriated tungsten, silicon bronze filler, 40A.

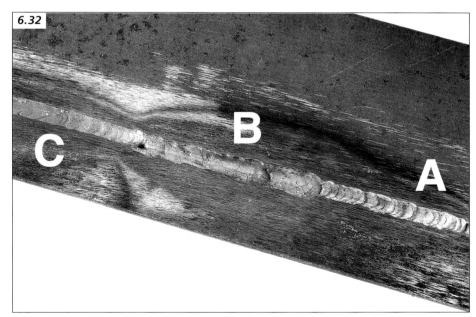

6.32. A lap joint TIG-brazed at the right current and forward speed runs quiet and smooth (A), with virtually no melt of the parent plates. Deliberately slowing travel increases heat input to the point that the underlying plates fuse with the bronze filler, which promptly cracks (B) and loses strength. Speeding up again resumes normal service (C).

Welding technique is straightforward. Use a slim thoriated tungsten, argon shield gas and just enough current to heat the joint quickly to red. Feed the bronze filler towards the arc and let it melt on the hot plates at the fringe of the arc; work quickly, using a melt, flow and move-forward sequence. Depending on the joint type and welding current, some very limited fusion of the parent plates is OK. If the joint gets too hot you'll soon know — the bronze will spread itself thin and wide, the underlying plate will melt and dilute the filler and the work is likely to crack (6.32). Too cold and the bronze will be sluggish and reluctant to flow.

So there it is. TIG can be the most frustrating of process, but is highly satisfying when you get it right. Minority sport or not, it's a very worthwhile addition to the workshop.

Section 7

Plasma Cutting

Plasma cutter: There's a name with a real *Star Wars* ring to it, even though the technology dates back to the 1950s and isn't nearly as sci-fi as it sounds. But it's still very clever stuff — and since the recent introduction of small inverter-based sets, plasma cutting is increasingly affordable on the farm.

What can it do for you? Traditional gas cutting kit works by oxidizing hot metal, and there lies its Achilles heel. Faced with something that won't oxidize (stainless steel, aluminum) or oxidizes poorly (cast iron), oxy-acetylene or oxy-propane can't cope. A plasma cutter, though, can slice anything which conducts electricity — so all the steels and non-ferrous metals are suddenly fair game (7.1). Using plasma takes less skill to use than oxy-fuel cutting, is potentially safer (there are no fuel gases and no flashback risk) and fillets metals with speed and surgical tidiness.

Yet it's not the universal solution for agriculture. Today's sets are still very expensive compared with gas cutting kit and consumables don't last long. Most single-phase machines have limited capability and are tied to a compressed air supply, so aren't readily mobile — though there are notable exceptions. Neither can plasma heat or weld. But if you have a need to make speedy, neat cuts in single items or stacks — and don't want the hassle and cost of cylinders — then plasma is definitely worth a look. For an idea of how it works, see page 112. Before then, here are two quick definitions: a **clean or drop cut** has a clean finish with minimal, easily removed dross and is produced by a plasma set working within its capability. A **severed cut** will separate but needs much more cleaning up, and tells you that the set is literally working on its ragged edge. The difference between the two can be just 0.08"-⅛" (2-3mm) plate thickness (7.2).

Needs, Options, Costs

A plasma machine will fit right into most workshops. All that's required is a single-phase mains outlet for smaller sets (13A is usually OK if the plug has a slow-blow fuse) or three-phase power for bigger ones, plus a reliable supply of clean compressed air. The latter may not be needed as some small plasma cutters have a compressor on board, though generally this means lower capability as the mains must feed both the compressor and the welding circuit. Around 4-5cfm (120 l/min) air at 80-90psi (5.5-6bar) will serve single-phase sets; something deliverable by even a modest compressor. Three-phase

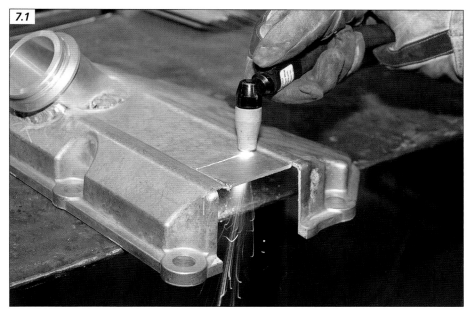

7.1. Gas can't slice aluminum or stainless steel — but plasma can.

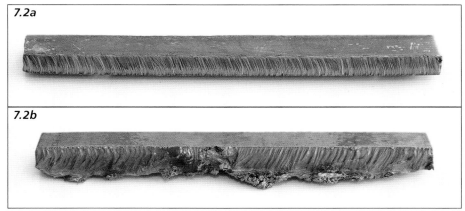

7.2. Only a small increase in thickness turns a quality drop cut into a sever cut. Here it's the difference between ⁵⁄₁₆" (8mm) plate (top) and ⅜" (10mm) plate (bottom).

7.3

7.3. A good single-phase inverter set has stepless control of current, warnings for over-temperature and low air pressure and a regulator/gauge for pressure setting. The air check button simulates flow to allow accurate setting. (PHOTO COURTESY OF ESAB NORTH AMERICA)

Main plasma consumables are the electrode and tip. An hour's continuous work before replacement is normal; new electrodes are around $10-$12, tips $4-$6. Outer nozzles ($19-$25) wear slowly but are at risk of physical damage. The cost of consumables is offset for many users by plasma's speed, ease, and ability to work round a template — which comes in very handy when making weathervanes and such.

How Does It Work?

A plasma is gas that's been superheated until it breaks down and ionizes. Plasmas are intensely hot (up to 90,032°F but typically 27,032-36,032°F [up to 50,000°C but typically 15,000-20,000°C]) and conduct electricity very readily.

Equipment varies from highly complex, mechanized industrial units to portable units (7.4). All work on the same principle. Simple single-phase sets make their plasma from ordinary air; three-phase sets use air too, but may also add a second gas such as nitrogen.

The plasma is generated inside a torch. Smaller and handier than the gas cutting equivalent, in its simplest form this holds a central electrode, a tip with a small exit hole centered over the electrode, and a surrounding ceramic/fiber cup (7.5). When the operator clicks the torch trigger, compressed air at 4-6bar (60-90psi) streams down around the electrode and exits through the tip.

After a short pause the power supply delivers a jolt of high frequency DC to the electrode, creating an internal pilot arc between the electrode and tip. This ionizes the air stream into a plasma, which pushes the pilot arc out from the tip. On contact with the work the set's main arc fires up and tracks down the plasma. The pilot arc is shut off, and the main arc/plasma forms a straight, clean-edged column travelling at near-supersonic speed. The combination of its force and heat first melts, then blows away metal to leave a narrow, clean cut line (or kerf) with very little dross. The cut continues for as long as there

tackle able to sever 1¾" (45mm) mild steel needs around 5.5-6.0cfm (165 l/min) at similar pressures.

The supplied air must be dry, as wet air leads to uncontrolled arcing in the torch which will quickly wreck it. Good sets have their own inlet air filter/regulator to back up the workshop outlet's own separator, and both of these must be serviced regularly.

Plasma can also pierce and gouge, albeit within limits. Piercing capability extends to around 60% of cutting capacity, while gouging is largely the preserve of three-phase sets.

The market divides into inverters (small, portable) and conventional power units. Either way the cost/capability balance strikes hard. Most single-phase equipment is limited to drop-cutting ⅜" (10mm) steel, which on most farms won't be enough for plasma to be the

only tool. One notable exception comes from Hypertherm, whose 45 can drop-cut 1" (25mm) mild steel and also gouge.

Three-phase sets are significantly more expensive. So as usual, equipment choice depends on the depth of your pocket and planned use. Where light-gauge work like ducting, irrigation pipes, cladding, trailer sides, vehicle bodywork, gates, guards and decorative ironwork feature, a small single phase plasma cutter with ¼" (6mm) capability ($600-$800) brings convenience without big capital outlay. From there you can chose between big-name brands, pay up to $2,500 and cut up to ⅝" (16mm) mild steel: maybe more. Beyond that come three phase models at upwards of $4,500 with capacities beyond 1¾" (45mm). Below top tier equipment lies is a swarm of smaller fry, much of it made in Italy, priced lower and with variable performance.

7.4. Dense technology inside a modern inverter helps explain the price.

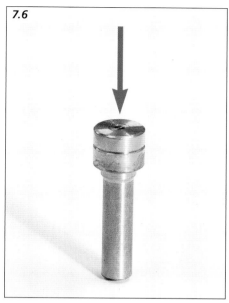

7.6. A plasma electrode has a central core of hafnium or tungsten (arrow), surrounded by copper. Service life is around one hour's continuous use, cost is $5-$6.

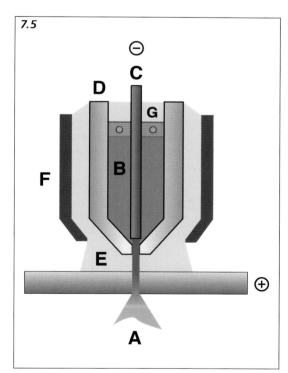

7.5. Torch operation. An intensely hot plasma jet (A) is formed from swirling high-pressure air (B) as it's ionized by an arc from the central electrode (C). The nozzle (D) concentrates the jet through a small hole. Cooling air (E) flows round the nozzle, contained by a shied cup (F). A ceramic collar around the electrode controls air swirl (purple, G).

is metal for the arc to transfer to, or until the operator releases the trigger. Air then continues to flow for a few seconds to cool the torch.

By varying the current, gas flow, gas type and tip hole diameter, the size and force of the plasma jet can be changed. Different thicknesses of varying materials can thus be cut. By using lower gas flow in dedicated kit, welding is possible — though not with the sort of equipment we're dealing with here.

Air is deliberately swirled into a mini-tornado as it enters the electrode chamber. The spin centers it on the electrode and forms the blast into a tight, parallel-sided column exiting the tip.

The electrode itself is has a core of hafnium or tungsten sheathed in copper (7.6). Hafnium (a close relative of the zirconium used in TIG electrodes) is used with air as it can stand contact with

oxygen; tungsten is used with nitrogen. The tip is also copper and would normally melt from proximity to the intensely hot plasma, but is protected by a skin of non-ionized, cooler air formed in the swirl process.

Torches vary in design. Single-gas versions (naturally enough) use one gas, typically air, but divide it into two flows internally; one swirled for the plasma and another passing around the outside of the tip to cool it. This shield flow also boosts cut quality, reduces spatter and helps when piercing. Multi-gas torches allow different gases to be used for plasma generation and shielding. Torches on bigger kit are usually single-gas and cooled by water.

Setup

The torch has an on-off trigger and in some cases, a stepped tip designed to follow a guide or template: the latter is useful when working in thin material (7.7). When precision is needed in thicker stuff, run the outer nozzle (rather than the tip) along a guide.

Torch components are few and fit together logically (7.8). Switch off mains

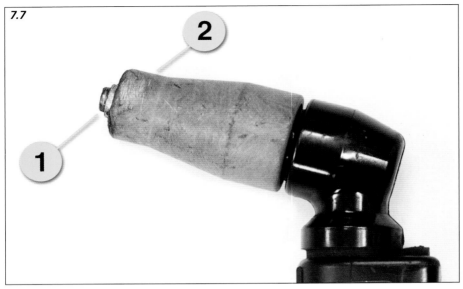

7.7. Some tips are stepped (1) so they can follow a sheet template. In thicker metals, rest the nozzle (2) on a guide if required — see 7.12 for an example.

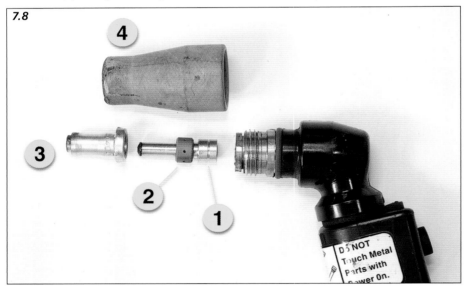

7.8. Torch components are the electrode (1), ceramic swirl baffle (2), tip (3) and nozzle (4). Electrode and tip are consumables.

7.9. Compare a new tip (front, center) with worn ones. As the central hole loses size and shape, cut quality goes down.

power before delving into the innards. Don't lose the ceramic swirl baffle: if this is not in the right place or is damaged, plasma won't form properly and arcing inside the torch is likely.

The electrode and tip are both consumable. Work with an electrode until it's about 0.2" (5mm) shorter than a new one: beyond this, the pilot arc will be reluctant to jump from the electrode to the tip and the torch won't fire up.

The tip itself both squeezes and directs the plasma jet. Cutting performance will suffer as the central hole erodes. With proper setup and a good tip, the cut line or kerf will be clean, narrow — much narrower than with a gas torch — and square-shouldered. Replace a tip as soon as the center hole elongates or grows (7.9), or when the kerf widens and becomes untidy. The outer ceramic/fiber nozzle will gradually deteriorate but unless whole segments break away, this won't affect cutting performance.

Plant Setting

External or internal, the compressor must be able to deliver ample air without flagging, or cutting performance will go rapidly downhill. Around 120 l/min air (4-5cfm) should serve single-phase sets, while bigger ones need 165 l/min (5.5-6.0cfm). Fit the air supply point with a water trap and turn off or remove any air tool lubricator. if a short extension hose is needed to get the set to the job, use at least 5⁄16" (8mm) bore.

A good plasma cutting set also has its own inbuilt water trap/regulator (7.10). This, and the one on the supply outlet, must be empty and clean. Adjust air pressure to the maker's requirement on the set's own regulator, using the air check facility (where provided) to mimic cutting conditions. Typical pressure will be 5.5-6bar (80-90psi).

All that's left is to dial in the amperage required for a clean, full-depth cut in the material you're working with. Trial and error is the way forward until you've worked out what your set can do.

7.10. Good sets have their own regulator/water trap.

Stay Safe

Plasma cutting is an arc process, so it's vital take the usual precautions against skin burns and eye damage. Cover up exposed areas and use a welding filter specified for arc use. Gas welding goggles will not do. Arc intensity is relatively low, so shades EW5-EW10 handle all work.

In this Murex torch, the business end is isolated from the power supply by a small switch button (arrow), released as the outer cup is unscrewed.

High open circuit voltages (200V-plus) at the torch add an extra hazard. It follows that torch maintenance must never be carried out with the plasma cutter switched on. Better-designed torches have automatic backup protection in the form of an internal switch: in the Murex version pictured, a cut-out button (arrow) isolates the torch from its power supply as the nozzle is unscrewed. Such backup is only there to cover the odd time that you make a mistake. Don't rely on it: switch the set off at the mains.

The thermal conductivities of aluminum and stainless steel are different to carbon- and cast steels, so cutting capacity is lower in these metals. Naturally, thickness also has an effect: using high current on thin material won't cause problems as long as you speed up forward travel to suit, while in thick stuff you'll have to slow down so the cut can penetrate. The set's capacity limit is reached when the kerf won't clear and slag blows back towards the torch.

Technique

A caution or two to start. When slicing conventionally with a gas torch, even a tiny air gap stops the cut — but not so with plasma. The positive news is that you can stack-cut several sheets at once. The negative is that when you're trying to separate one thing from another (say, a damaged bearing from its housing) you can no longer rely on the gap between parts to stop the cut. On top of that you'll need to take more care when working over something you might not want damaged, like an anvil.

The return lead needs a clean, zero-resistance attachment point. Grind a clear spot for it otherwise performance will be poor. As with any arc process, if current passes through the small contact points of bearing balls or through bushings on its way back to the set, resistance heating is likely to flat-spot them.

Cuts in thin materials can be made with the torch tip resting on the work. For higher currents and thicker materials, leave a 0.08"-⅛" (2-3mm) stand-off gap between tip and work. A specially designed stand-off nozzle can be used where consistent top-quality results matter (7.11).

To start a cut, position yourself where you'll be able to see the tip all the while.

7.11. A stand-off nozzle helps make consistent cuts in thicker material, but limits the operator's view. A good metal or composite shield should run easily alongside a guide without arcing.

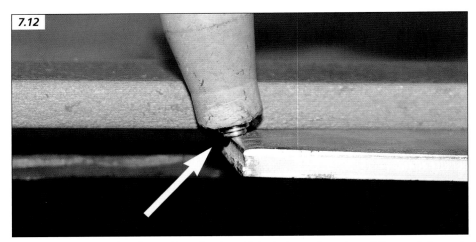

7.12. To start a cut cleanly, position the tip with its hole on the edge of the work (arrow). Working with sheet, hold the torch upright. With plate, angle it as shown — this is ⁵⁄₁₆" (8mm) mild steel.

7.13. The right travel speed sees a continuous stream of molten metal clearing the cut. Dross (arrow) will chip away easily.

7.14. Compare torch angle with the molten flow leaving the plate to see how the plasma column exit point lags. Note the very narrow kerf width and lack of distortion — both are characteristic of plasma cutting.

Hold the torch so the tip's central hole is just in contact with the work (7.12): this allows the plasma stream to initiate quickly. Then squeeze the trigger. After a burst of compressed air the arc fires up and the cut starts. Move off at a speed which keeps a steady stream of molten slag flowing from below (7.13, 7.14).

On thin sheet hold the torch vertical. In thicker stuff the plasma's exit point will lag behind the entry point, so allow for this by angling the torch slightly forward in the direction of travel (7.14). At the end of a cut, keep the trigger down until the sliced plate falls away — if you let up too soon, the plasma stream won't clear the bottom edge and a tiny island will be left uncut.

Making Holes, Gouging

Plasma can pierce holes in any conductive material, though maximum depth is limited to about 60% of the set's cutting thickness capability.

As when starting to pierce with gas, molten metal will splash up from the surface. If fine metal spray interferes with the magnetic field around the plasma column, the column can sway into the tip and damage it. Or if molten metal bridges between the tip and plate, high current will flow through this bridge and wreck the tip completely.

Get round both problems by starting to pierce with the torch canted over at roughly 45°. Once the initial splash subsides and a crater opens up, gradually bring the torch upright. Maintain a standoff gap of about twice the normal cutting height between tip and plate, or as much as the set will allow. Work round the embryo hole until it's the required size.

Unless you're blessed with a good eye and a steady hand — or better still, a circle cutting attachment — some modification is usually needed to tidy up the hole. While plasma lets you carve slices from the edges to round out a hole or to increase its diameter, this is not done as easily as with gas. The plasma

Plasma vs. Oxyacetylene

Here's a thumbnail comparison between cutting with a capable single-phase plasma set and with oxyacetylene. Factors below will be influenced by the volume and type of work on any given farm, and the care taken over plant setup and use.

Factor	Plasma	Oxy-acetylene
Initial cost[1]	High	Modest
Mobility[2]	Requires compressor	High
Consumable life	Medium-short	Long
Consumable cost	Medium-high	Low
Plant complexity	High	Low
Range of material	Wide	Narrow
Distortion	Low	High
Ability to heat	None or limited	Excellent
Ability to weld	None	Good
Ease of gouging	High	Moderate
Stack cutting	Easy	Difficult
Max cut thickness	Limited	High
Potential cut speed	Medium-fast	Medium-slow
Cut quality	Excellent	Variable
Post-cut cleanup	Little/none	Medium/much
Skill required	Little	Moderate

1 Assumes adequate compressor is already on farm.

2 Sets with an on-board compressor are mobile but capacity may be limited; by how much depends on brand and model.

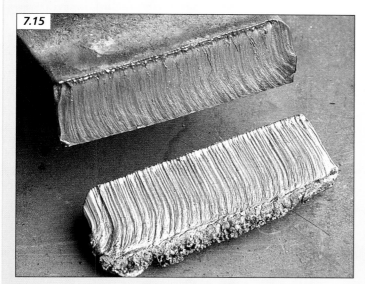

7.15. Single-phase operation need not mean restricted cutting capacity. 1³⁄₁₆" (30mm)-thick mild steel plate sever-cut by a 240V Hypertherm Powermax 45.

7.16. As the plasma stream bridges small air gaps, stack cutting is possible – here's two bits of ⅜" (10mm)-thick mild steel.

stream stops immediately when there is no conductive metal below it.

Gouging is a good way to cut a preparation U-groove before welding. Plasma is ideal for the job, but usually you'll need a three-phase set — most single-phase units don't have the grunt to produce a long plasma column. Which is a pity, as gouging with plasma is faster, more effective and less fume-ridden than with a MMA rod. Use a wide tip, hold the torch at around 40° and blow metal away, repeating if necessary until the right profile/depth is reached.

Problems, Causes

Here are some common faults and their fixes.

- **Reluctant arc strike.** Poor earth return connection/path; worn electrode too short; air pressure too high.

- **Substandard cut.** Material too thick for set's capability; electrode too short; tip orifice misshapen or too big; travel too fast; poor earth.

- **Heavy dross build-up.** Travel too slow; current too high; material thickness at set's upper capability limit.

- **Material distortion.** Travel too slow; current too high.

- **Internal arcing.** Ceramic anti-swirl baffle missing or damaged, water in compressed air supply.

- **Fluffy-looking plasma column.** Water in compressed air supply.

- **Reduced cutting speed/thickness.** Low air pressure or flow; current set too low.

Bonus Section

DIY Welding Projects

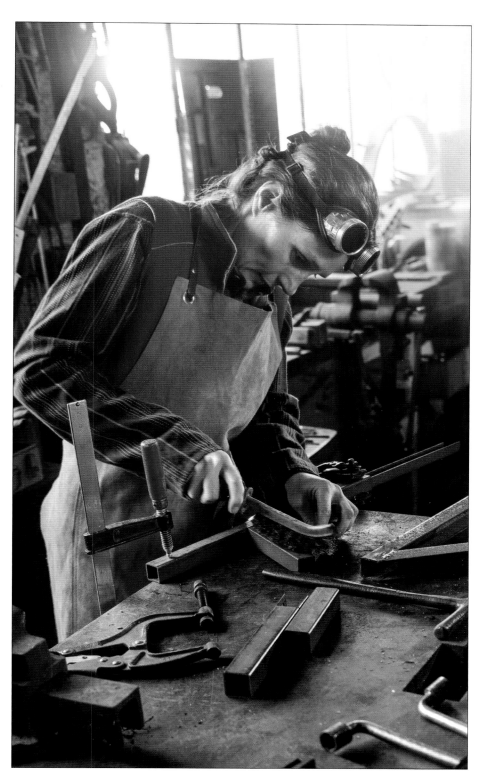

The point of learning welding and metalworking skills, of course, is to make things from metal (and also to repair them). The following chapter contains examples of what this can look like. The intention is to show you the workflow of a welding project.

Special thanks to Forney Industries and Hobart Welding, two venerable welding equipment manufacturers who graciously provided the material for these demonstration projects. They represent some creative thinking and good workmanship. The expectation is that welders who are reviewing them will use them to spur their own creativity and help develop their skills and workmanship.

Project Prep

Always remember to practice safe metalworking habits by wearing the proper personal protective equipment. A fire-resistant welding jacket or apron and sleeves, welding gloves, safety glasses and a welding helmet are vital to protect yourself against dangerous arc flashes and sparks.

Welding Cart

BY RACHEL BIGUM
COURTESY OF FORNEY INDUSTRIES
WWW.ASKFORNEY.COM

If you have all the welding gear but no cart to stow it on, you've come to the right place. Joe Mooney from Homesteadonomics ran into the same problem after he got his new Forney MP welder. Follow these step-by-step instructions and you can create your very own three-shelf welding cart.

TOOLS AND MATERIALS

- **Scrap steel**
- **Magnets**
- **Welder**
- **Sheet metal**
- **Galvanized steel**
- **Expanded metal**
- **Wheels (two sizes)**

STEP BY STEP ASSEMBLY

Begin by tacking together the lower shelf. This shelf will eventually hold the wheels and the uprights that will hold the upper two shelves.

Using tack welds and your magnets, begin forming the skeleton of the cart. Use the magnets to hold up the second and third shelves, and tack them in.

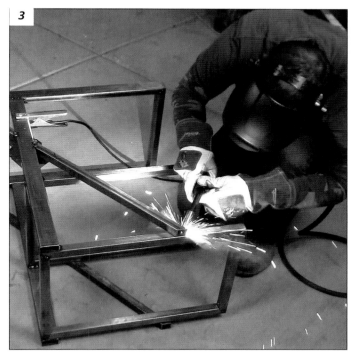

Create a diagonal support from recessed metal. This diagonal support can serve two main purposes: to give the cart stability and support, and also to improve the appearance. Tubing is shown here, but you can also use flat scrap pieces or a gusset triangle.

From the same recessed metal, create and tack-weld cross supports for the shelves. This step is to ensure you can put sheet metal on the shelves.

After you finish the mainframe, you can begin creating a toolbox out of galvanized steel for the second shelf of the cart. This box can also act as a drawer. It is not generally recommended to weld with galvanized metal due to the toxic fumes it releases. It's better to use regular sheet steel and paint it later.

Now, grab your expanded metal and begin creating the bottom of drawer/tool box. You will be adding wheels to make it a sliding option.

Attach the bigger wheels to the back end of your cart to make it mobile.

Add the smaller wheels to the front end of your cart. With two sets of wheels your cart and your machines can travel with ease.

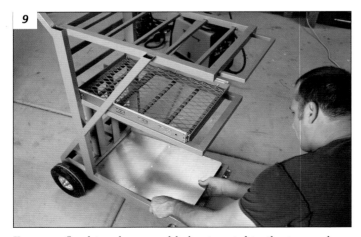

For your final touch-ups, add sheet metal to the top and bottom shelves of the cart for a more refined look.

Optional Accessories

If desired, you can add fun and functional features to your cart such as a hook for a welding helmet or clamps, a milk crate for extra storage, or a bottle bracket for your gas tank.

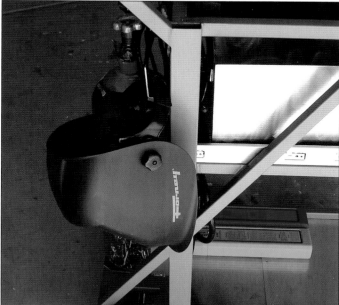

Welded Steel Table

BY BLAIR WEILNAU
COURTESY OF FORNEY INDUSTRIES
WWW.ASKFORNEY.COM

A welding machine offers endless possibilities to your DIY needs, without being a burden on your wallet. Does your broken lawnmower need repair? Do you need to fix your metal trellis or gate? How about build your own smoker or fire pit? Not only can you repair your broken equipment, but you also have the capability to design and create your own sculptures, garden art or furniture pieces that will last a lifetime.

In addition to the basics, every welder needs a welding table to help them get started. It's an essential piece of being a successful welder and offers many benefits. A metal table allows you to easily ground your work and is sturdy enough to handle the toughest welding job. Plus, it's easy to make on your own. We decided our table would be 3-feet tall, by 4-feet long and 2-feet wide.

TOOLS & MATERIALS

- 190 MP welding machine with .030" MIG wire
- 700 P plasma cutter or a reciprocating saw
- Expanded metal
- 2" wide and ¼" thick steel (angle iron and flat iron)
- ¼" thick steel plate
- Angle grinder with grinding and cutting discs
- Metal cutting chop saw
- Heavy-duty magnets
- C-clamps
- Welding helmet, safety glasses, welding gloves, and a fire resistant jacket/apron.

STEP BY STEP ASSEMBLY

The first step is to cut all angle iron, flat iron, and expanded metal to your desired length. We started with the frame of the tabletop. Using a plasma cutter, we cut angle iron into two 4-foot pieces for the length and two 2-foot pieces for the width. Next, we cut the angle iron into four 3-foot pieces for the legs.

PRO-TIP:

Use a clamp or heavy-duty magnet to help hold the expanded metal in place while you make your cuts.

Cut the expanded metal to be used for the bottom shelf. We started with a 24" x 24" piece and cut it in half. Each 12" x 24" piece will fit a quarter of the bottom shelf and will be supported with flat iron. Using an angle grinder with a cut-off wheel, cut through the expanded metal.

3

On the two 2-foot pieces of angle iron for the table top frame, cut out a section of the corners so the 4-foot pieces of angle iron can rest inside of it. You will eventually weld these pieces together. We used a reciprocating saw to cut these small pieces.

Grind Small Parts to Fit

Grind down the edges that don't match up; you want them to fit flush together. Remember to square up these pieces and ensure they create a 90° angle; this will help the table maintain its rectangular shape. If you notice that the edges don't meet exactly, grind them down rather than re-cutting. Grinding is easier and will help you get the snug fit you're looking for. If you have a large piece that needs to be removed, revert back to your plasma cutter or reciprocating saw.

4

Arrange your pieces on the ground and make sure everything lines up and fits together well. We measured the diagonals at this point to make sure the top was square and even, as this is the foundation for the rest of the table.

5

Cut the flat steel pieces for the shelf. One piece should be 4-feet in length and there should be two 1-foot pieces. We used a chop saw with a 14" cutting wheel to cut the flat steel.

6

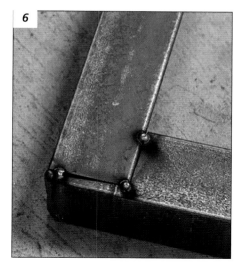

To begin welding, start by welding the frame of the top piece together. Ground your clamp to the workpiece. This ensures the welder is ready to strike an arc. Use tack welds to secure your pieces together. Tack welding temporarily holds the pieces together. Start by welding along the edges/corners. Once you confirm the frame is square and even, seal up the rest of the gaps.

7

Next, weld the legs on to the frame (four 3-foot pieces of angle iron). We put the tabletop frame on top of a flat surface and used heavy-duty magnets to keep the legs in place. Use the magnets to your advantage and secure it the best you can. If you don't have heavy-duty magnets, use c-clamps. Complete one leg at a time and don't try to secure all legs at once.

8

We used a level to make sure the legs sat flush to the frame and used tack welds to secure the legs in place. Weld the legs in place after leveling and tack welding is complete.

9

Next, add a bottom shelf to hold welders, tools, cords, etc. The shelf will be made from expanded metal and the shelf frame will be made from angle iron. The cut angle iron measures 2" wide by 20" long. We used C-clamps to hold the angle iron nine inches from the bottom. We leveled the pieces and tack welded them to the legs. If your shelf isn't level, instead of taking off the clamps and repositioning, try using a hammer to move the metal up or down until it's where you need.

10

After securing the shelf frame with tack welds, move the table to the floor and begin welding. Moving the table to the floor will give you more space to complete your welds. Add support cross bars for the shelf with flat steel (one 4-foot piece and two 1-foot pieces). Use a C-clamp to hold them in place, tack weld each side and secure the pieces to the shelf frame.

Take each 12"x 24" piece of expanded metal and spot weld it to the angle iron and flat iron pieces. A spot weld is more secure than a tack weld and is sufficient enough to fuse metal together for prolonged use. Spot welding will also help prevent burn through. Burn through is caused by applying too much heat and burning a hole or gap in the metal. If you try to run a long bead, the expanded metal will get too hot. Spot weld the expanded metal on the bottom side of the table for a cleaner look.

We wanted to add a 90° angle below the tabletop frame so we can hang our angle grinders. We measured one 14" piece and one 8" piece of 2" wide flat steel, clamped them in place with a C-clamp and welded them to the angle iron.

Finally, get ready to assemble the _" steel plate to the top frame. You're going to clean and prep the steel plate with denatured alcohol to ensure a cleaner, more structurally sound weld. Using an angle grinder with a grinding wheel, grind down the welds that are on top of the frame to ensure the steel plate will sit flush.

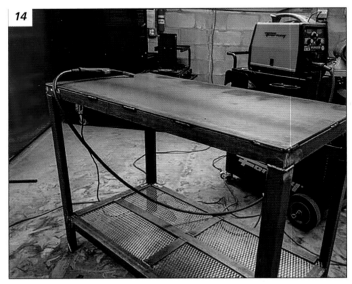

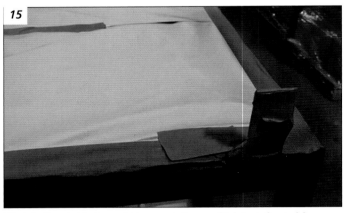

Finish the table by adding paint or polishing the table top to get the customized look you desire. The final result will be a working, sturdy table that will not only hold and store your welder and welding supplies, but assist you with future welding projects.

Line up the steel plate and fuse it to the frame with 3-inch welds across the edges. Doing smaller 3-inch welds will keep the table aligned and prevent warping. If you do a continuous line, the heat will warp the table and you will end up with an uneven base. You can leave the 3-inch welds as is or go back and fill in the gaps. You can also grind down the edges or leave them to display your great work.

PRO-TIP:

Add a post to the edge of your table so you can easily ground your future workpieces. If you decide to paint your table, remember do not paint the ground post. Painted metal will not ground.

Metal Fire Pit

COURTESY OF HOBART WELDING PRODUCTS HOBARTWELDERS.COM

A flared fire bowl made from welded steel sheets is captured within ladder-style sides to create a pleasing contemporary fire pit that is also portable. The careful workmanship and thorough grinding leave a smooth surface that is easy to clean.

A few notes of warning: obviously, a metal firepit is going to get very hot, so do keep children well away from it when the embers are burning. And be sure to check your local codes: most specify a minimum distance the outdoor fire pit must be from any permanent structure, and do beware of low hanging boughs, Finally, if you value your lawn, avoid setting the firepit up in the middle of a high visibility area: hot fires and grass do not get along well.

TOOLS & MATERIALS

- Plasma Cutter
- MIG Welder
- 12 – 1" × 2" × .065" square tubing
- 4 – 1" × 1" × .065" square tubing
- 8 – 1.5" × 1.5" × ⅛" angle iron
- 1 – 4' × 4' × ³⁄₁₆" steel sheet

OPTIONAL TOOLS

- Scrap metal
- Grinder
- Clamps
- Paint

MATERIAL DIMENSIONS

- Top Frame: 4—1.5" × 1.5" x .125" angle iron x 36" L (cut ends at 45 degrees)
- Legs: 4—1.5" x 1.5" × .125" angle iron × 18" L
- Side Vertical Supports: 4—1" × 1" x .060" tubing x 33.5" L
- Side Bottom Supports: 9—1" × 2" × .060" tubing x 8.5" L
- Bottom Leg Supports: 4—3" × 3" × ³⁄₁₆" flat stock (optional if fire pit place on soft surface)
- Pit Sides: See drawing

STEP BY STEP ASSEMBLY

Cut the four ³⁄₁₆-thick by 3 × 3" leg support pieces from the flat steel. Tip: Use a straight scrap piece of metal to help create straight line.

Cut the tubing and angle iron to length. Then, cut a 45-degree angles in top four edges of angle iron to make mitered corners.

Use an angle grinder to bevel and smooth the mitered edges of the angle iron.

Clamp two of the angle iron corners down to welding table and make sure the mitered joints are square.

5

Tack-weld the clamped corners together with your welder.

6

When all four corners are tack-welded, flip the frame over and make sure you are squared. Adjust as necessary. Measuring the diagonals is a good way to check for square.

7

MIG-weld the weld the top of each corner, wearing appropriate safety gear.

8

Use an angle grinder to clean up and smooth out the joint. Do careful work here.

9

Tack the legs into corners on the inside of the frame so the corner welds will not be visible.

10

Begin building the fire box by placing the bottom square on the table, and then tack-weld each side of the box. Tip: When placing the sides of the fire pit together you are going to want some assistance holding up the sides.

With the frame and the firebox upside-down on the table, place the firebox inside the frame.

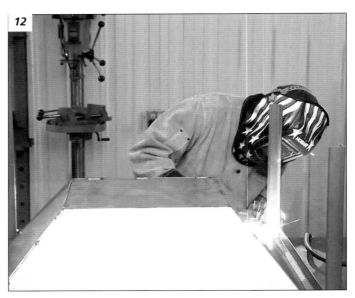

Tack, square, and weld the firebox to the frame.

Weld the decorative "ladder" side pieces by placing 1" × 1" × .065" tubing horizontally on the table with three pieces of 1" x 2" × .065" tubing standing on top. Weld the parts together and clean up with a grinder.

Turn fire pit box on its side so it is easier to weld it together with the sides.

Deluxe Dog Waterer

COURTESY OF FORNEY INDUSTRIES WWW.ASKFORNEY.COM

Are you tired of constantly filling up that water bowl for your thirsty four-legged friend? Make your life easier by using these step-by-step instructions to build an eye-catching dog water tower. Forney Industries and Joe Mooney from Homesteadonomics teamed up to weld a water tower fit for a canine king or queen.

TOOLS AND MATERIALS

- **Scrap steel**
- **16-gauge galvanized sheet metal**
- **Magnets**
- **Water barrel with desired dimensions**
- **Shelf brackets to match the angle of your chosen barrel**
- **Galvanized float bowl and attachable hose**
- **4½" angle grinder**
- **Flap discs**
- **Plasma cutter**
- **MP Welder**

The float and tube system used here works kind of like a toilet tank, filling the water bowl automatically when the float gets too low.

STEP BY STEP ASSEMBLY

Begin your project by assembling your base frame of the tower, then tack weld the frames together using your desired dimensions.

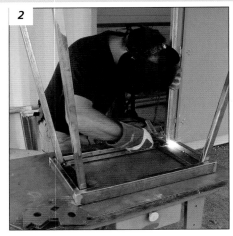

Tack weld the legs onto the metal inserts. Dimensions: one 2" 16 ga for the top, one 1" 16 ga for the legs, ¾" 16 ga for the leg supports. After you've finished your welds, begin cutting some recessed metal that would act as the diagonal support beams.

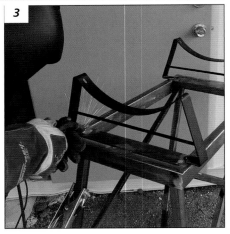

Now that the water tower is roughly assembled, weld your shelf brackets to the edges of the frame. These shelf brackets should match the angle and curvature of your chosen water barrel size. The brackets shown are old barbecue grill brackets. Brackets can be fabricated or found materials, all you have to do is repurpose them!

Make your tower look nice and polished by cleaning off the surface rust with your 4½" Angle Grinder and flap disc.

Cut some metal strips from 16-gauge galvanized sheet metal. These strips will act as straps to secure the barrel to the tower. A plasma cutter makes quick work of cutting the straps. At the bottom of your tower, attach the galvanized float bowl and hose.

Set the water barrel on top of the beams and secure the straps into place. Now that you are finished assembling the tower, you can either paint it or leave it as is.

Section 8

Special Techniques

1. Cast Iron Welding

Back in Granddad's time, cast iron and farm machinery were inseparable. These days there doesn't seem quite so much of it about, although now and again a broken sprocket or some such is likely to thump on the bench for repair. When it does, what are the welding options?

One thing is for sure — it's funny old stuff, as likely to crack as not and never coming with a guarantee of a successful fix. And the dirtier the casting, the more unpredictable is the outcome. Welding problems stem from three main areas:

- **Contraction stresses.** Cast iron is brittle, and often can't put up with different rates of contraction between the weld and parent material. So just when you thought you'd got away with it, the work goes "ping" and cracks, usually in the heat-affected zone alongside the weld. Back to the drawing board.

- **Item shape.** Castings usually have nooks and crannies, changes of section and complicated shape, all of which make good preparation difficult and accommodating those contraction stresses harder.

- **Oil.** Absorbed into the material's micropores and held on graphite in its structure, oil vaporizes under welding heat to leave weakening porosity in the joint.

Just to muddy the waters a bit more there are several sorts of cast iron, some of which can be welded and some that can't. To sort out which is which, see box "What Have You Got?" and use the tests suggested. And older castings in particular often have high levels of sulfur and phosphorus, neither of which help when it comes to welding. Later ones are at least better in this respect.

Assuming the defunct chunk turns out to be gray cast and thus potentially weldable, there are two ways to go about it. Or five, though option 3 (fusion welding with oxyacetylene) involves thick filler rods and big flames, and options 4 and 5 (MIG/MAG and TIG) fall more in specialist territory. On most farms the likely options are bronze welding with a gas flame and filler rod, or stick welding with the arc set.

CLEAN UP FIRST

Cast iron repair is not necessarily cost-effective, as the Box "Is Welding Worth It?" shows. But where the sums do add up or repair turns out to be the only option, the first thing is to minimize the chance of the job cracking. Take a close look. Are there existing cracks? These are very likely to spread through cast iron's coarse grain structure as the joint cools and pulls, so stop any you find by drilling both ends to full depth (8.2).

8.1. With the right technique and consumables, welding cast iron need not be (too) scary.

Next comes **cleaning,** which is definitely something not to be skimped. As with any weld, start by grinding away dirt and paint. Clear an area about ¾" (20mm) on either side of the joint, going down through the casting's hard outer skin.

Gearbox casings or similar are likely to be steeped in oil, which over time worms it way deep through the metal's tiny interconnecting pores or latches on to its graphite flakes. While the ideal solution is to hold the whole thing at 752°F-932°F (400°C-500°C) for several hours to burn oil out, your kitchen oven probably doesn't go beyond 482°F (250°C).

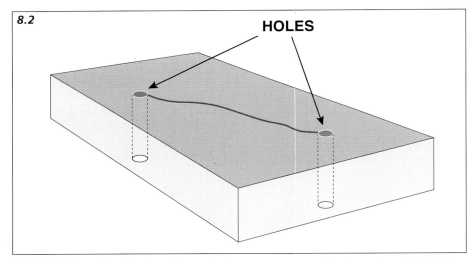

HOLES

8.2. Cracks spread easily through cast iron's coarse structure. Look very hard to find the exact ends of a crack, then drill ⅛" (3mm) holes to full depth to stop it.

The best plan is let nature take its course. If possible, first steam clean or hit the parts with strong solvent, wearing appropriate personal protection as required. Weld, then grind out the porous layer and repeat the process until weld metal is solid. Each time you grind back, leave a thin "buttering" layer

of new metal on the parent plates. If you're fusion welding, cast iron's high carbon content will be diluted progressively into the weld so the joint ends up stronger.

Joint preparation is a little different to normal. As always the idea is to open up the area to allow full-depth penetration, leaving a small step at the root so parts can fit accurately back together. But grind to produce a U-groove rather than a Vee, sloping the sides to about 30° and rounding off sharp edges to minimize

heat concentration (8.3). If you have them, gouging electrodes are good at producing the right shape and the arc heat helps clear contamination. If the joint faces show dirt-holding blowholes, drill these clean.

GET HOT UNDER THE COLLAR

With the job prepared, think about preheating. You may get away with welding the job at room temperature, or not; it's unpredictable. Warming the work first always helps. But how hot is hot, and do you need to heat the whole thing?

Like most of us humans, cast iron hates nasty shocks. So heating and cooling both need to be slow, and the slower the better. Ideally, all but ductile castings are best brought evenly up to 932°F (500°C) and then cooled very gradually so stresses can release. That's fine in industry but not on the farm, where the aim must be more modest.

Preheating is achievable with a gas torch and will help all weldable castings. After cleaning and U-preparing, lay the work on a hearth built from old firebricks, walling it in as much as possible. Then — keeping the torch on the move — use a cutting nozzle to heat the whole job if it's small enough, or as much of it as you can manage. At the

8.3

30°

A

80°-90°

8.3. Grinding or gouging a U-groove is the best way to prepare cast iron, though a wider-than-usual Vee is OK. Leave a small step (max ⅛" [3mm] deep) so the jigsaw can fit back together accurately. Rounding off all sharp edges prevents local hotspots which would otherwise overheat bronze filler metal.

8.4. A Tempilstik melts at a defined temperature, showing when a casting is up to preheat. For castings, 482°F (250°C) is a good target, easily achievable over a substantial area with a neutral gas flame.

very least, heat enough to evaporate all surface water. Otherwise bring the job to 482°F (250ºC) (low black heat), and if possible up to black red heat (around 896°F [480º C], visible only in low light). To make the process more certain invest in a pair of Tempilstiks (8.4), waxy crayons which melt at an exact temperature when rubbed or pre-marked on a hot surface: welding suppliers have them. Once the job is up to heat, weld immediately.

GAS WELDING GRAY CAST

The less heat that goes into the weld zone, the less likely is cracking. So bronze welding — a relatively cool process — looks promising. The joint faces must be COMPLETELY clean or the bronze can't "wet" them, so thorough grinding is vital. Round off sharp edges which will otherwise overheat and oxidize, stopping the bronze from keying.

Rather than welding the joint in one go, butter one joint half at a time. That is, lay down a thin coat of bronze on each. Use the normal bronze welding technique (pages 87-88). Settle on a nozzle which brings the work quickly to red heat but allows control over the flow of filler. To weld, use plenty of cast-compatible flux (check the packaging), a slightly oxidizing flame and a manganese bronze filler rod. Be sure to use **manganese** filler — this has a higher bond strength than the more common silicon bronze, so produces a stronger joint. It also wets cast iron better.

Flow filler metal over one face (8.5). Work as fast as possible to minimize heat build up, but slow enough to be certain that wetting is complete.

What if bronze won't wet an area? Grind it clean, reflux and try again until the buttering layer is complete. To finish the job bring both halves together and weld, re-melting the buttering layers as you go and adding more flux as required. Multiple runs will fill a deep joint, but keep heating to a minimum.

Tips for Success

- Weld cracks in the middle of items from the center outwards. If a crack runs to an edge, weld from the fixed end to the edge.

- Leave a small (⅛" [3mm] max) step on each broken face to allow parts to be fitted back together before welding. Thick sections can be ground out from both sides to produce a double-U preparation.

- If you can't pre-heat, weld in short sections using a backstep technique (page 85). Let each weld cool to touchable temperature before starting the next.

- Think about cutting out a cracked area and replacing it with a mild steel plate. This often works with thin castings. MMA or bronze welding will handle dissimilar metals, but check the rod specification first. A plate around half the thickness of the casting it replaces is usually plenty strong enough.

- Where the repair is highly stressed, it may be worth bolting a mild steel brace over the area to spread the load. Things always break at the point that stress concentrates!

- With thin sections, minimize weld metal deposit (and thus contraction stress) by welding in the vertical down position where possible. Check the electrode for suitability first, though, and try it on a bit of scrap to settle current and technique.

ARC WELDING GRAY CAST

Rod choice is critical. All makers produce specific high-nickel electrodes, with ESAB's OK 92.58 good. This iron-nickel mix handles various cast irons, can be used in all positions including vertical down, and its low OCV requirement (50V) suits it to any good AC set. DC set owners might try running the electrode DC positive: the arc will have more force and the electrode will

8.5. A silicon bronze filler rod wouldn't wet this ½" (12mm) gray cast plate (right), even with generous fluxing. But the preferred manganese bronze rod would (left). After grinding off excess filler the buttered plate could be laid alongside another, and the joint made by re-melting both buttering layers while building up the Vee with bronze.

8.6. On your marks. The starting positions for nozzle and filler rod: travel will be from left to right. Note that nozzle angle is steeper than for normal bronze welding. Dipping the heated rod into general-purpose flux produces a generous coating. Wiping flux into the joint helps too.

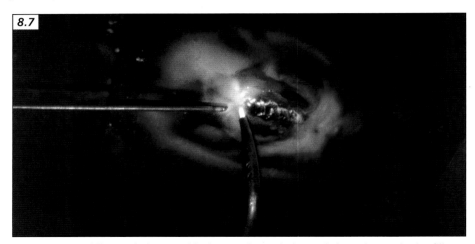

8.7. Bronze welding technique. With the nozzle angled steeply into the work, the filler rod is dipped into the molten pool of bronze and withdrawn.

8.8. When arc welding cast iron, break a long run into short sections. Weld one, let it cool, weld the next. Backstep in thick sections and to minimize heat input, don't weave a root run.

perform a little better. If the rods are damp dry them at 392°F (200°C) for 2 hours, oven permitting.

As with gas welding, minimum heat input is the game. Use the smallest diameter rod and lowest current you can get away with. Don't let the electrode's typical soft, fluffy arc tempt you into turning up the wick — stick to recommended settings.

Clamp the joint halves as loosely as possible, then weld. In thick sections or on cold joints, keep heat input local by breaking the first run into 2" (50mm) sections and not weaving. On long joints space these first welds well apart, then backstep between them to fill in the gaps. On heavy section needing multiple passes it's a good ruse to butter the joint faces first. Not only does the buttering layer reduce dilution from below, but it absorbs contraction stresses from subsequent runs.

AFTER WELDING

With cast irons, what happens after welding is just as important as what went before. Two things minimize the chance of cracking; very slow cooling, and for arc welds, hammer peening.

8.9. Peening the hot weld with a blunt chipping hammer does wonders for stress in the weld (and in the welder).

What Have You Got?

The following recaps information on pages 9-10. Castings are given away by their complex shape, rough surface finish and any raised lettering. To tell cast iron from cast steel, grind a sample. If the sparks are yellow and bushy, it's a steel. If sparks are dull orange and short, it's cast iron (8.10 and 8.11).

Separate the different irons by application. Anything used in a high-abrasion situation is probably white or chilled cast iron. This is very brittle, with fresh- broken faces showing silvery white with very coarse grains. White cast iron is not weldable.

The most common stuff is gray cast iron, which turns up as around 70% of castings and holds its 2%-5% carbon content as graphite flakes. These bring some lubricating properties (so gray cast is used in machine beds, pistons and such) and give a fractured face its color, but they make excellent planes of weakness. Grey cast iron is weldable with the right electrode or filler rod and careful post-weld cooling. Typical applications are belt pulleys, bearing housings, electric motor mountings, machine tool beds, exhaust manifolds.

Where a cast part needs to be straightenable or has to handle high

8.10. Cast steel produces plenty of bushy yellow sparks under the grinder . . .

8.11. . . . while with cast iron it's more a question of spot the spark, which will be fewer, shorter and redder.

shock loads, it's likely to be malleable cast iron. Produced by heat treating white cast so its graphite flakes turn into less fracture-assisting aggregates, malleable cast is weldable but may lose resilience if heated too much or cooled too quickly. Uses are gearbox housings, bench vices, cutter bar fingers and so on.

Peening releases stress in the weld zone and should be done **straight after each run, while the weld is hot.** Tap the surface all over with a blunt chipping hammer, keeping it at an angle to avoid generating cracks (8.9).

Slow cooling is vital no matter how the joint was welded. Think of the host casting as an old lady with hemorrhoids; keep it away from cold surfaces and draughts, and don't even think about plunging it in cold water. The best

plan is to leave the repair among the firebricks of your temporary hearth, cover it with dry sand or a fire blanket, and don't go near it for a couple of hours. At least that way you won't hear it crack.

2. Pipe Welding

Next to vertical work, joining pipes or other circular things is the job most likely to give welders the willies. The problem is truly three-dimensional: the target area curves, making it tricky to hold a reasonably constant angle between the electrode/torch and weld pool. Keeping the angles right is tricky when you're joining a vertical pipe to a flat plate; harder still when welding round a horizontal pipe, where the working position must slide seamlessly from vertical to down-hand, back to vertical and then to overhead. Doing it with a cow nibbling your overalls is downright improbable.

In production work the pipe is often rotated slowly past the operator in a bringing-the-mountain-to-Mohammed exercise. But in the middle of a windy stockyard, Mohammed has to get himself to the mountain. The easiest way is to divide the joint into segments, then weld each separately (8.12).

Two joints — a butt and a fillet — show how to crack the main problems. In each case the basic strategy is the same: divide and rule. The solutions work for both MMA and MIG, though MIG's lack

of slag and good angle-tolerance make it easier.

FIRST STEPS

The pre-welding sequence is the same as for any job. Prepare the joint, choose the rod or wire, set the plant, and tack.

Preparation. As always, what you put in determines what you get out. At the very least the joint area has to be ground free of rust and rubbish. In non-critical applications, pipes can be just butted together and welded. Strength will then be confined to the bridging weld, but for gates and such this is enough.

Where strength matters more, Vee-grind the joint faces (8.13). For the best chance of controllable full-depth penetration and maximum strength, leave a ⅛" (3mm) vertical root face at the pipe end(s). Strip the flux from a ⅛" (3.25mm) rod, snip off a section, bend into a U-shape and slip it between the joint halves as a spacer which can be pulled clear after tacking (8.14).

Electrode/filler wire choice. Any good quality, general purpose MMA electrode or wire is fine where vertical segments will be welded upwards. Check that information on the electrode

pack confirms rod suitability for vertical up and overhead work. ESAB's 46.00 is a reasonable choice, along with equivalents in other brands.

If you have trouble welding vertically up, work down. In material over ⅛" (3mm) thick there are two choices. Either leave a good penetration gap (vital for MIG) make multiple runs, or pick a specific

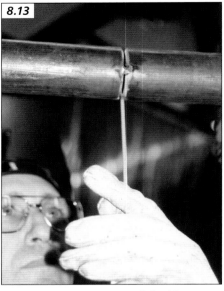

8.13. For maximum strength in this ¼" (6mm)-wall pipe, preparation includes Veeing the ends and leaving a ⅛" (3mm) penetration gap.

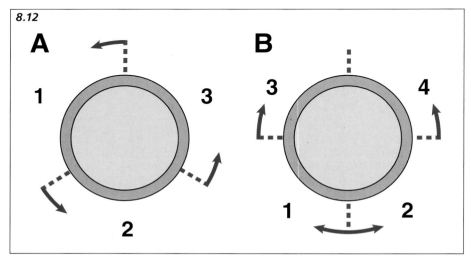

8.12. Divide and rule. Split pipe joints into segments, each small enough to weld in a continuous pass. Seen from above, a fillet joint (as in pipe to plate) is welded sequentially, here in three segments (A). Seen end-on, a butt joint between pipes (B) is usually split into four segments: welding usually starts from the bottom (runs 1 and 2) and finishes at the top (runs 3 and 4). The actual order and direction of welding depends on access and electrode. Start and stop each segment away from tacks.

8.14. MMA core wire, bent into a U-shape, makes a quick gapping tool.

pipe electrode. Penetration is potentially poor coming vertically down with a general-purpose MMA rod, but pipe welding versions like Arcos Lincoln's Nufive are designed specifically for the job. Check the set's available open circuit voltage (OCV) against the rod's requirement before buying to, as they say, avoid disappointment.

For MIG work up or down, grade ER70S-6 filler wire is fine. Team it ideally with Argon/20% CO_2 shield gas for maximum penetration in thicker material.

Plant setting depends on the joint.
For butt welds in pipe using MMA and welding vertical segments upwards, set current towards the bottom end of the rod pack's recommendation to give maximum control of the weld pool. Try a run on scrap; if sticking or slag traps are a problem, increase current a little. If welding vertical segments down, set higher current for best penetration and slag control. For fillets welded on the flat around a vertical pipe, use current towards the top of the recommended range to help penetration and minimize traps.

The same principles go for MIG/MAG. On vertical work, set welding conditions

8.15. Small tacks are OK if fusion is complete — leaves less to weld over.

(voltage/wire speed) which give a steady arc at slightly less current than you would use on the flat. Where weld metal build-up is a problem, cut back on wire speed and travel faster — but do neither to the point where arc stability and penetration take a hit. On fillets, use higher voltages. The aim is always good edge fusion and a flat or slightly convex bead.

Tacks. These need to be fused into the work, which suggests using current on the high side of normal (8.15). Where joint integrity will really matter, grind off the leading and tailing edges of tacks to give a tapered ramp; this allows better fusion with the main weld metal. Start and stop away from tacks, welding over them as you go.

A few other thoughts. Comfort is the first: try to support your own weight (and the mass of the electrode holder or torch) by draping everything over any handy support. Keep your arms free to move. Use a welding helmet, rather than a hand mask, to leave digits available for steadying a torch or guiding an electrode. Before you start, make a dummy run to decide how far round the joint you can comfortably reach and see. During the actual run, stop as soon as control gets difficult, when you can't see the weld area, or if you simply keel over through holding your breath too long.

DIVIDE AND RULE

With the work prepared and tacked, lay out your strategy. Mentally split any circular joint into three or more segments, with number decided by access and joint diameter (8.12).

Four divisions are normal on larger items. Butt welds in horizontal pipe are generally split into four quadrants, each starting and finishing on a main compass point, that is, north, south, east and west. If it helps, chalk start and stop points on the work.

Where the job can be welded in the flat (down-hand) position and turned between welds, do it that way. If it's

fixed, tackle each segment in turn and reposition yourself between runs.

When producing fillet welds (like joining a pipe to a plate, or pipes at 90° to each other), consider fillet thickness. For maximum strength, fillet depth should equal the thinnest joint component. As a rule of thumb, the following electrodes produce the following fillets from a single pass:

- ³⁄₃₂" (2.5mm) rod leaves a ⅛" (3mm) fillet
- ⅛" (3.25mm) rod leaves a ⁵⁄₃₂"-0.2" (4mm-5mm) fillet
- ⁵⁄₃₂" (4.0mm) rod leaves a ¼" (6mm) fillet

Where single-pass welding is not possible (either through lack of the right rod or enough current to run it) use multi-runs to build the necessary fillet depth. Multi-runs are also useful where a fillet joint has been Vee-prepared, as a big-diameter rod is unlikely to reach right down into the Vee. In such a case, fire in the first run with a ³⁄₃₂" (2.5mm) electrode, then build up fillet depth with bigger items.

The sequence and direction of welding is up to the operator, with the limits set by access, electrode type and individual ability. If you're using MIG it's worth setting the nozzle back from the contact tip so you can monitor stick-out and get the torch close into the joint. Too much stick-out lowers welding current; ⅜" (10mm) is maximum. On vertical segments current needs to be low anyway for best weld pool control, so watch stickout doesn't grow or fusion will suffer.

BUTT JOINT, PIPE LAID HORIZONTAL

After dividing the job into four segments, start at the very bottom — six o'clock (8.16). From here weld upwards to nine o'clock on one side, then up to three o'clock on the other. For these first two runs, keep the electrode as near 90° to the work as possible, hold a short arc (with MMA) and don't weave. If a penetration gap has been left, maintain an "onion" shape above the weld pool and listen for the hollow crackle which confirms full penetration. If burn-through starts, weave a little to re-direct heat into the pipe walls. Steady the rod or torch with your free hand, as every little bit of control helps.

With the first two runs in place, the next two meet at the top. During these passes you'll find it hard to rotate your arm fast enough to keep ahead of the rod tip. Depending on how tolerant the electrode is of odd angles, it may be OK to finish with the tip actually trailing the weld pool — but don't make slag traps. If you don't like welding upwards it's OK to start at the top and weld down to meet the first runs, but only if material thickness and preparation allow full penetration.

With all four runs in place, grind out any lumpy starts or humps where tacks have been welded over. To fill a Vee repeat the sequence, only this time weaving gently and pausing at the edges to melt them in. Finally seal it with a capping run, sticking to a diagonal weave to keep things simple. Otherwise it's like trying to ride a bike while juggling eggs.

FILLET JOINT TO HORIZONTAL PLATE, PIPE VERTICAL

Joining pipe to a flat plate is a little easier. Either butt joint components together, or for more strength, taper-grind the pipe end and tack to leave a penetration gap. Set MMA current towards top of recommended range to minimize the chance of slag traps. A little undercut on the pipe is OK if it will be filled by following runs.

Be sure you can see round all the intended segment before starting, and that your welding arm and the cable or torch are free to move. Start each segment forward of any existing weld, then fill the existing run's finish crater before moving off.

For strength, fillet depth should equal the thinnest joint component. So when joining (say) ¼" (6mm) wall pipe to ½" (12mm) plate, look to end up with a ¼" (6mm) fillet. A ³⁄₃₂" (3.25mm) rod leaves a ⁵⁄₃₂"-0.2" (4-5mm) fillet, so a capping run will make up the shortfall. If you're weaving this use a simple diagonal pattern, pausing at each edge to allow full fusion. Three un-weaved overlapping runs with a ³⁄₃₂" (2.5mm) rod will produce the same result and be marginally stronger, but will take longer. It's up to you.

8.16. Start at the bottom. Weld upwards to 9 o'clock on one side, then the other.

8.17. The finished job: neat, strong, no leaks.

3. Hardfacing and Gouging

HARDFACING

Hardfacing is something of a mystery. It's like driving round the farm on a foggy morning — all the familiar landmarks are there but somehow different, and you're not necessarily sure that you're looking at a profitable operation.

Still, this much is clear. Hardfacing is the business of laying down material that resists abrasion (and perhaps impact) on new or worn mechanical or soil-engaging parts, with the aim of extending their service life.

WHICH WELDING PROCESS?

The speed and ease of MIG/MAG welding suggests that's the way to go. But hardfacing with MIG means using cored wires, and to get the required burn-off and fusion with these needs more current than single-phase sets can deliver. So for many farms, the weapon of necessity is a manual metal arc (MMA) stick set. Depending on the rods you choose even that may not be up to scratch, as the box "Get Set and Go" explains (page 148).

Matching electrode and job takes us back into the fog. Most repair and fabrication involves mild or low-alloy steels, so you can jog along quite happily with general-purpose rods; their forgiving composition handles surface contamination and minor variations in metal make-up.

But wear-resistant parts are made of sterner stuff. In the workshop it's impossible to fathom their composition, so the repairer can't properly match electrode and material in the way that industry can. The wrong rod and/or technique can see expensive hard covering flaking off, and perhaps even the whole part cracking through — although with a little care neither should happen. To make sure it doesn't, here's the low-down on electrode selection and care, plant setting, pre- and post-heat treatment and welding technique.

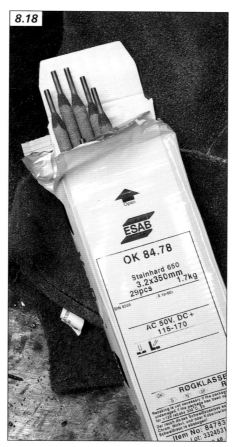

8.18. ESAB's hardfacing consumables come vacuum-packed. Plant setting and re-drying requirements are on the pack.

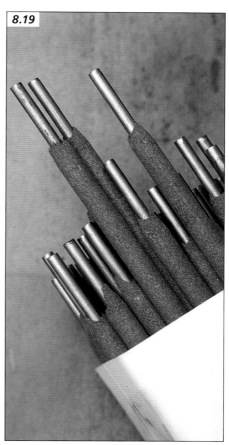

8.19. The rod's thick coat contains metal powder, so hardfacing consumables lay down more material than core wire diameter alone suggests. In other words, the recovery rate is more than 100%.

First though, please take two cautions to heart — see "WARNING: Manganese and Fume" (page 149).

ROD SELECTION AND CARE

Hardfacing rods can be formed as a tube packed with flux and powdered metal, or as a conventional coated stick. ESAB offers the latter (8.18 and 8.19).

Four ESAB electrodes cover farming's needs:

• **OK 83.28** is a low-alloy rod designed for building up parts in metal-to-metal contact: gears, sprockets, dog clutches and so on. Hardness is 30 HRC, machinability is good. Min OCV 70V, AC or DC+ operation.

• **OK 83.50** is a low alloy, high carbon electrode for repair and build-up of worn parts. Runs from small sets with low open circuit voltage. Resistant to abrasion and impact. Hardness is 50-60 HRC; can only be ground, not machined. Min OCV 45V, AC or DC+ operation.

• **OK 84.78** contains chromium carbides and is for surfacing only. Ideal for very abrasive soils or corrosive conditions. High recovery rate — that is, it puts down a lot of metal for a given rod diameter. Hardness 59-63 HRC so can only be ground. Min OCV 50V, AC or DC+ operation.

• **OK 84.84** is a very expensive, complex carbide-containing rod for surfacing only, suited to extremely abrasive soils where pressure on the part is high. Good for edges, unusual in giving high hardness from a single layer. Hardness is 62 HRC, can only be ground. Min OCV 45V, AC or DC+ operation. Keep rod vertical to work while welding.

Notes: Equivalent rods are available in other brands. HRC is a comparative hardness scale.

The electrodes above come vacuum-packed to keep them dry. Once opened and stored, all four types should be oven-baked before use, or water in the coating is likely to produce hydrogen cracking in the weld. Re-drying instructions are on the packet, along with info on welding current range, OCV requirement and so on (8.18).

TO BUFFER OR NOT?

The harder the deposit, the less flexible and more brittle it is. If very hard material is laid directly over substantially softer metal, the top layer may crack on cooling and can flake or spall off altogether, and anyway your expensive hard metal is diluted by mixing with the softer stuff.

One way round cracking and spalling is to use a buffer or buttering layer between soft and hard metal, like the jam in a sandwich. ESAB's OK 67.45 lays down the required tough, stress-absorbing buffer, while dilution can be kept down by welding relatively quickly, and by using minimum current and minimum weave to restrict melt and mix.

How do you know when to butter? This is formally defined according to parent material and impact level in service, but lacking accurate information on the former you'll have to suck it and see. The simple way is to hardface a few trial parts without buttering, and if the top coat stays put in work you'll have the answer. The more complicated — but much better — approach is to badger the maker of the part to be surfaced for info on its composition, then to ask an electrode supplier about the best technique and rods for the job.

8.20. When pre-heating, keep the torch up from the work and on the move to spread heat evenly. After hardfacing, cool slowly in air.

8.21. Use conventional arc length, travel speed and rod angles when hardfacing. For overlapping passes, angles are roughly 60° from vertical, 45° from parent plate.

Importantly and irrespective of buttering, at least two hardfacing layers are usually needed for the top one to show maximum hardness.

PREHEATING AND POST-WELD COOLING

While not all hardfacing rods require it, pre-heating and slow cooling of the work make good insurance against cracking. Having said that, some particularly abrasion-resistant hard facing materials will crack anyway, usually with no ill effects in service — its info sheet should cover that.

Usually pre-heat is to a specific temperature, but in the rough-and-ready workshop it's enough to warm the part evenly and thoroughly with a gas torch (8.20). The aim is just to drive off any water, so don't overdo it. After welding let the work cool slowly in air. Quenching is a sure way to produce cracks and risks embrittlement of the whole part.

TECHNIQUE

Nothing out of the ordinary here in rod angles, speed of travel, arc length and plant setting — just keep within the requirements laid out on the packet and weld as you would with a general-purpose rod (8.21). Expect a fluffy arc punctuated by quiet spittings (much like cast iron or special-steel rods) with a lacy slag covering to follow. Watch your eyes with this, as bits can ping off spontaneously as the weld cools.

Hardfacing electrodes are designed to give a quick-freezing deposit which helps when working along edges, but paradoxically the weld pool tends to be wide. This makes working away from the flat rather tricky; use the shortest arc, lowest current and fastest travel consistent with good fusion to control the pool. Overhead work is best avoided or left to owners of a thoroughly fireproof hat.

WEAVING

There is only one rule concerning hardfacing weave patterns, and it's that there are no rules. How (and when) you do it depends on the wear that you're trying to repair or to stave off, and how much money you like to spend on electrodes.

Having said that, the more and wider you weave the more heat goes into the work; the more diluted the expensive hardfacing metal becomes, and the more likely you are to end up with slag trapped in the deposit.

Mega-heat input is never a good thing when welding, producing extra distortion and often a greater chance of weld failure though cracking, particularly in the heat affected zone alongside the bead(s). And high heat input from over-enthusiastic weaving is also very bad news if you must hardface high manganese steels or brittle cast iron.

Some surfacing metal dilution is inevitable, and can even work in your favor when laying softer material on high-alloy steels. But normally the

Control your weaving to make best use of hardfacing rods.

hardfacing layer is thinned by the softer base metal that you're trying to protect, so in general you'll need to put down two or even three layers of hardfacing to end up with the material's full benefit. Either way, proper current selection and minimum weave is a priority.

The rule of thumb when weaving is simple: go no wider than 3x the electrode's core wire diameter. That'd be ⅜" (10mm) tops for a ⅛" (3.2mm) rod, and so on. In fact, thanks to hardfacing rods' high recovery rate you can cover a lot of ground with a relatively narrow weave pattern, so the restriction isn't so bad.

TAKE YOUR PICK

Although there are as many patterns as people laying them down, a few principles will see you right. Clearly it makes sense to fully protect the leading edge of a soil-engaging part, usually with overlapping runs (8.22).

Side faces of thick items are often best tackled by laying down parallel beads, one along each edge, then filling the gap between them with a simple side-side weave (8.23). Depending on rod diameter, sides can also be covered by a side-side weave alone (8.24); in both cases the rod's fast-freezing deposit helps preserve the shape of the edge.

During the weave, pause briefly at the end of each sideways movement to let the crater fill. If the build-up in the center of the weave is too heavy, either travel faster across that section or move a little further forward each time.

Away from tips and sides, exactly where you put the hardfacing is set by the likely wear pattern. A worn part is the best guide here, showing the areas where pressure from soil or metal-to-metal engagement scallops away the steel. When facing new parts, it's not always necessary to deposit great swathes of hardfacing. Instead, a criss-cross pattern (or even many large dots) creates a grid which fills with soil (8.25). Earth then flows over this so can't reach the soft base metal below. Bucket vertical sides and chisel plough points are good recipients of criss-cross weaving.

GOUGING

How do you cut grooves or holes, Vee — prepare joints or slice away old metal and rusted parts without a grinder, a gas torch or a plasma cutter? With a gouging/piercing electrode.

ESAB lists only one — OK 21.03 — so rod choice is as simple as the technique. As the rod's coating burns it develops a very strong gas jet, and this blows away metal melted by arc heat. OK 21.03 works with steel, stainless steel, cast iron and most other metals apart from pure copper, and is particularly good for weld preparation in cast iron; arc heat dries out and burns off graphite and contaminants which would otherwise increase cracking risk. It's also handy for opening a groove on the reverse side of a welded joint, prior to laying in a sealing run.

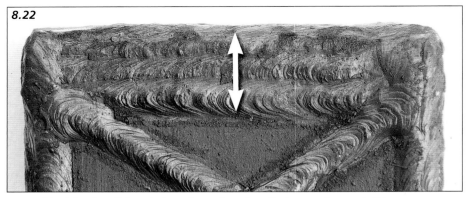

8.22. This slab of 1" (25mm) plate shows how hardfacing can be used. If it were a soil engaging part the leading edge (arrow) would take most wear, so is faced with overlapping runs. A second layer would counter dilution.

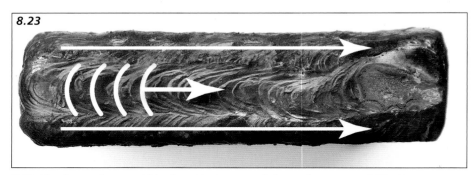

8.23. The long edges could be faced by two parallel runs (long arrows) filled by a compact weave . . .

8.24. . . . or depending on electrode diameter and part thickness, by a single weave.

8.25. Away from maximum pressure points, a criss-cross weave hold soil to protect the underlying soft metal.

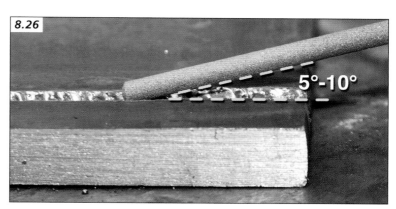

8.26. Strike a gouging rod vertically, then quickly lay it at a 5°-10° angle. As soon as the gas jet starts to dig, work the rod forwards and back like a saw.

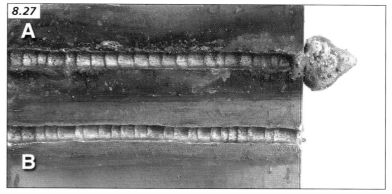

8.27. The rod pack suggests current setting and gives OCV requirement. When both are right there will be a little slag (A), which chips off easily to leave a clean, ripple-bottomed channel (B).

There are a couple of mini-snags. OK 21.03 needs 70 OCV minimum, so can't be used with most small air-cooled sets. When cutting and piercing it prefers to be run DC+ (that is, with the electrode holder plugged in to the positive line); so if the farm has only an AC set you won't get the best from it. The grooves and holes you make won't be millimeter-precise, and by it's very nature, a gouging electrode can never be used to join material.

TECHNIQUE

A gouging rod is used like a saw. To open a groove, strike the arc with the rod vertical to the work then quickly lay it down, ending up almost parallel to the surface — an angle of 5°-10° is good (8.26). As the jet starts to open a trough, work it forwards and back as if you were cutting with a hacksaw.

The gouged channel will show a series of ripples corresponding to your sawing, like sand on a beach when the tide has gone out (8.27). To cut a deep groove make one run, clear any slag, then repeat as necessary.

Stick to rod pack suggestions for current. When this is right the gouged channel will be largely clean and any slag will chip away easily. If the cut won't clear and fills up with foamy clinging slag, try more amps. If it's blasted into a wide channel and fills with tenacious slag, reduce current. Figure 8.28 has unpleasant examples.

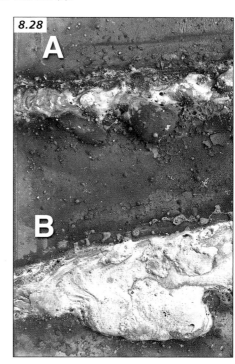

8.28. Too little current limits gas jet size, so it can't clear tenacious slag (A). Too much current sees the gas jet overwhelmed by slag from the wide channel cut by arc heat (B). Either way it's a mess.

Gouging can be done on the flat, vertically (start at the top and work down) or horizontally. Overhead work isn't advisable unless you are a complete masochist.

PIERCING AND CUTTING

Quick'n'dirty holes can be chopped out or made bigger with a gouging rod. Though crude, the technique throws a lifeline to workshops without alternatives.

Strike the arc with the rod vertical, then once the gas jet is established, saw the rod down through the plate (8.29). As with gas cutting, the first stage until breakthrough is the hardest bit.

Once the arc is bellowing through the bottom of the plate, work round the edges to enlarge the embryo hole. Arc-pierced holes aren't likely to win many beauty prizes, but the technique is a good way to make big openings in heavy sections when more sophisticated, neater methods aren't around. It's also handy for rough-cutting away broken parts or seized fixings.

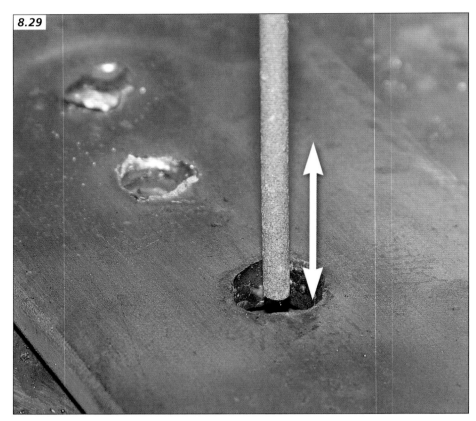

8.29. To pierce or enlarge a hole, strike the arc with the rod vertical. Saw it down through the plate (arrow) then round the edges.

Get Set and Go

A voltage — the open circuit voltage or OCV — must exist between the electrode and work before the arc can strike. Simple stick units have one output terminal and generally offer 50V OCV. Others provide a second line at up to 90V.

Stick welding electrodes always have a specific OCV requirement. General-purpose rods are usually happy to strike and run at 50V, but specialist rods—including some hardfacing varieties—need higher voltage to strike cleanly and run with a stable arc. So before buying consumables, ask the supplier for their OCV requirement and check that your set can deliver the goods. If it can't the results will be poor.

WARNINGS: Manganese and Fume

Don't hardface 12%–14% manganese steel. High-manganese steels are typically used in crushing and milling applications and, crucially, are not magnetic — something you can easily check if in doubt. Otherwise they're light gray in color and are usually castings, designed to be bolted on to get round welding them. Sometimes they come with a warning label, and may have the letters "MN" cast into them.

Often used for bucket teeth and other bits which have to put up with high impact, high manganese steel is very tough but quickly turns brittle when reheated and slow-cooled — so the heat from hardfacing is very likely to crack it. If you must hardface or repair, weld very short lengths and keep the temperature below 392°F (200°C). Use minimum welding current, lay down a little material, let the area cool until you can touch it, then repeat. Work with several components at a time if you can, switching to another while one cools. Note: Structural steels with lower manganese content are fine.

Now to fume. Hardfacing electrodes carry a lot of metallic elements in their flux coating — which is how even a small diameter rod manages to lay down a very substantial bead. Consequently, hardfacing fume contains much human-damaging heavy metal vapor, which won't do you any good at all. Ideally use forced fume extraction. If that's not possible, wear a specific welding respirator marked EN149: 2001 (3M's disposable 06920 mask is an example). Work in moving air and stay out of the main fume column.

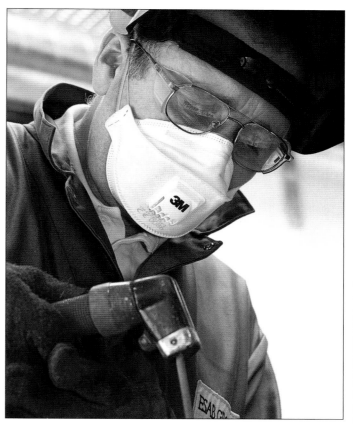

Hardfacing releases more hazardous metal vapor than general repair work. If fume extraction is not possible, wear a purpose-designed welding respirator and stay out of the main fume column. Nuisance dust respirators will not protect you!

Fumed Out

Gouging rods also produce mighty clouds of fume. Extraction is the best answer but failing that, wear respiratory protection and preferably work in an open doorway where the wind can do its bit. In the latter case, shield passers-by and livestock from eye-damaging arc light.

Health hazard. Gouging rods are best used with forced extraction.

Hardfacing prolongs active life for parts new and old.

Economics

Should you hardface or buy new? Or to put it another way, how long is a piece of string? It clearly depends on a three-part equation — your soil's abrasiveness, the cost of genuine or pattern bits, and the cost of the hardfacing consumables plus the labor to use them.

Naturally the equation balances differently for every farm. To help find where it settles for you, here are prices of the ESAB products discussed above. We've quoted costs without tax and on a kg basis, the latter to get round any difference in rods numbers per packet between different brands. Costs are recommended retail; your mileage will vary depending on distributor.

Electrode	Diameter	Cost $/lb (£/kg)
OK 83.28 Hardmat	1/8" (3.2mm)	36.31 (10.31)
	5/32" (4.0mm)	32.45 (9.21)
OK 83.50	1/8" (3.2mm)	57.98 (16.46)
	5/32" (4.0mm)	52.77 (14.98)
	3/16" (5.0mm)	46.67 (13.25)
OK 84.78	1/8" (3.2mm)	58.37 (16.57)
	5/32" (4.0mm)	56.65 (16.08)
OK 84.84	5/32" (4.0mm)	171.74 (48.74)
OK 67.45	1/8" (3.2mm) (others available)	73.81 (20.95)

Section 9

Soldering

Soldering is a lower-temperature version of brazing, using filler alloys that melt at lower temperature and are less physically strong. And as with bronze work, the bond between metals is not made by fusion. Instead it's formed partly by the filler hooking into the tiny hills and valleys of the joint metal surfaces, but mainly from the solder dissolving (not melting) into a very shallow surface layer of the joint. So for soldering to work, it must have free access to ultra-clean parent metals. For this reason flux is always used as a chemical backup to mechanical surface cleaning.

The process subdivides into two. **Silver soldering** suits dissimilar metals and is generally using with capillary joints, leaving little or no external buildup of filler. The filler itself is a fairly expensive mix of copper, zinc and silver which, being tougher than soft solder yet still electrically conductive, is useful where a joint must stand moderate heat and vibration. Electrical heating elements are often silver-soldered. The temperatures needed for silver soldering can only be achieved by a flame (or arc), not an iron. If the work is not too big it can be laid on a firebrick hearth and heated with a small butane torch.

The steps necessary for a sound job are the same as for brazing (pages 89). Buy the flux when you buy the silver solder. Take great care over preparation. Pre-flux the joint, then dip the rod in flux. Work quickly so the flame does not tarnish the cleaned surfaces. Let the silver solder melt from the heat of the joint metal, not directly by the flame. After soldering the flux residue can be washed off with hot water.

Soft soldering is the much more common form, used on heat-sensitive items or on joints that don't need much mechanical strength. Filler metal melts at 482°F (250°C) or below so the necessary heat can be transferred from an iron. The solder alloy is a lead-tin mix, often spiced with antimony. Soft-soldered joints are everywhere — in vehicle wiring, radiator header tanks, plumbing joints in copper pipe. Different metals can be joined too, like copper to steel.

Three things are necessary for soldering success:

- Clean parent metal
- The right temperature
- The right flux

So effectively it's brazing all over again. Cleaning is, as ever, a matter of elbow grease. Soft solder is even less tolerant of dirt than bronze, so it's essential to be thorough. Grind surfaces where possible — a clean flap wheel is good — then if necessary bring them to a shining finish with emery cloth. Dubious-looking or corroded surfaces must be scoured clean.

Fluxes protect the metal from oxidation and breaks down the liquid solder's surface tension to let it flow, boost wetting on the metals to be joined. **Paste** fluxes generally have a resin base and are relatively non-corrosive. They're used primarily for electrical work and come in a tin or as a core in the solder itself. **Acid** fluxes come in liquid or paste form, have a more vigorous cleaning action and are generally used on plumbing or sheet metal work. Being corrosive, they're not for delicate electrical stuff and should be washed from both the joint and your hands after use.

IRON OPTIONS

Heat for soft soldering comes from an iron. Why it's called an iron when the

9.1. Whenever an electronic component looks as if it could be cooked, grab its leads with pliers to siphon off soldering heat.

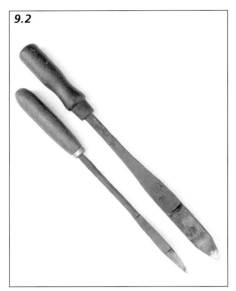

9.2. *Traditional soldering irons hold the bit in a steel shaft. Heat comes from a separate flame.*

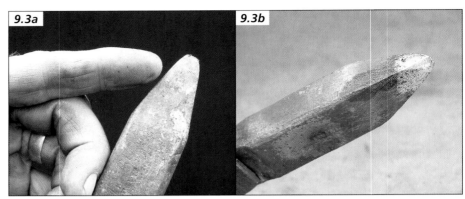

9.3. No matter how the bit is heated, it must be kept clean. The dirty-looking, oxide-covered specimen (a) won't transfer heat properly and brings contamination to the joint. File or wire brush it, then keep it wholesome during work by wiping on a damp cloth. A bit must be tinned before use; that is, fluxed, covered with a thin coat of fresh solder and refluxed to leave it shining silver (b).

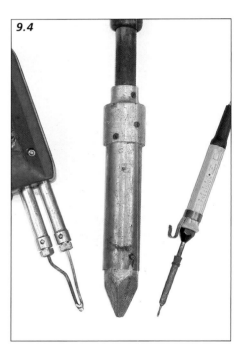

9.4. Three types of electrically heated soldering iron. The 20W miniature variety (right) is naturally for small jobs. The much bigger version (center) has a jumbo-sized tip, giving a good heat reservoir for big work. A 100W instant heat gun (left) is a good all-round tool, coming quickly to temperature and usually taking interchangeable tips for different work, including roughly fusing plastics.

business end is made from copper isn't clear, but there it is. The bigger the copper end (or bit), the greater is its heat reservoir, so the bigger the job that can be tackled.

The key thing is to match the iron to the job. If the iron is too small it won't put enough heat into the work and the solder won't flow. If it's too big, access may be difficult and solder will flow all over the shop. And while we're on the subject, electronic bits and bobs can be terminally damaged by heat — so whenever there's a risk of cooking a component, say on an alternator diode pack, use long-nosed pliers to soak up the therms (9.1).

The traditional soldering iron holds its copper bit in a steel/wood handle (9.2). The bit is generally heated by blowlamp or propane torch. If you're pushed a neutral oxyacetylene flame will do, but this must used with great care if the bit isn't to be overheated and covered by oxide scale (9.3a).

How do you know when a traditional iron is hot enough? Settle the flame over the bit's thickest section. Eventually the metal will flicker clear copper-colored and a green flame will ghost round it — now you're ready to boogie.

Irons heated by electricity or butane gas have all but displaced the old sort. Small versions are perfect for wiring and electronic work (9.4). Bigger ones can be backed by fast-acting heater coils, making an "instant heat" gun. At the top of the scale are jumbo electric irons, which take an age to get to full working temperature but supply plenty of heat for larger jobs. Self-contained, gas-heated irons let you solder anywhere and are particularly useful around vehicles and in a tractor cab (9.5).

Traditional or modern, any iron has to be cleaned and "tinned" — given a fresh coat of solder — before work. A tatty burnt tip won't transfer heat quickly and imports oxides to the joint. Titivate a dirty bit by wire-brushing or gently filing it back to bright copper metal while it's hot. Then tin the tip by sizzling it in flux and flowing on a little solder to form a bright silver jacket (9.3b). When you need to clean a hot tip during a job, wipe it on a damp cotton cloth.

MAKING THE CONNECTION

Whatever the work it's worth remembering that like all glues, a soldered joint is at its strongest where the bond layer is kept thin. If not, joint strength drops to that of the solder itself.

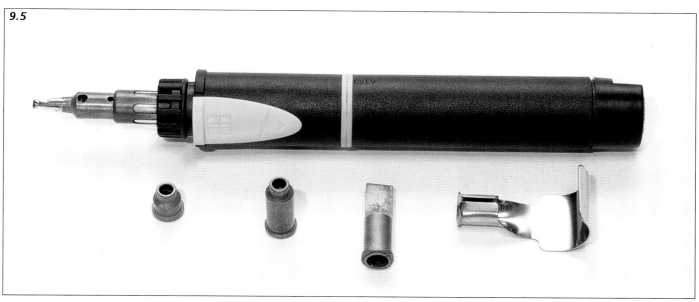

9.5. *Modern self-contained gas irons are ideal for jobs in the cab. A good brand like this Antex generates plenty of heat. Kit's attachments extend usefulness into jobs like cutting/sealing polyprop rope. Liquid butane is held in the handle/reservoir.*

Joint design is critical too — soldering works best where the joint has a large surface area, and where it's loaded in shear rather than peel. Picture 9.6 explains more.

We'll follow in pictures two jobs which show soldering's usefulness. First joining electrical wires, then fitting a cable nipple. To finish off, there's a quick look at soldering copper plumbing fittings.

Copper pipe fittings. Acid flux's better cleaning action helps with these. Over-the-counter Yorkshire fittings already have solder contained in the raised ring by each inlet. If more is needed (and it shouldn't be), use plumber's solder in stick form. Make sure this isn't covered by an oxide film; a scrub with emery cloth will shift that.

Instructions on the flux container usually say no joint preparation is needed, and usually it isn't. But to save the possible hassle of redoing the joint, whip round the pipe's end with emery cloth. That's particularly helpful if the fittings have been lying around the workshop for a while.

Wipe a little flux round the pipe end and slide the fitting on until it "bottoms." Using a propane torch, heat the fitting evenly round the raised ring where the solder lives. If you're indoors, a non-asbestos backing pad (buy from plumber's merchants) protects nearby plaster and paintwork. When everything is hot enough, the solder melts and is sucked through the joint by capillary action. Watch for it arriving as a bright ring where pipe and fitting meet.

Solder should have appeared right round the joint. If not, re-heat the area where it's missing and some should turn up. But if it doesn't? Reflux, reheat and add a little solder from the stick. If the work is hot and clean it'll be sucked into the joint. Take away the heat and hold the joint steady until the solder dulls.

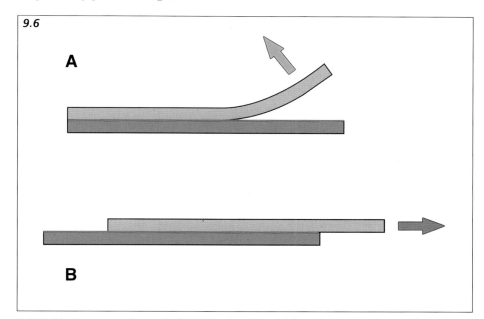

9.6. *Soldering uses soft metal alloy as a glue. Glued joints are weak when peeled (A) — think of pulling off a sticking plaster. Joints loaded in shear are much stronger (B). So a lasting soldered joint is one with a big surface area which will only have to cope with straight pulling loads.*

Soldering Copper Wire

Soldering cable can be anything from joining an electric fencer's leads to replacing a tractor battery terminal; the principles are the same. Every time you **tin the parts first, then join them by heating,** adding more solder if needed.

Electrical work needs a non-corrosive flux. Use flux-cored solder and back it up with a separate supply, like a tin of Fry's "Fluxite" (9.7). Proceed as follows.

Wire is only clean enough to solder if it's bright copper-colored. Where there's any sign of corrosion — as there will be in old trailer light cable or anything that seen much service and been exposed to the elements — tease out the strands and rub them between folded emery cloth until they shine. Twizzle the strands back into shape and dip the end in the flux pot.

Bring the iron up to temperature, which means a green flame around a traditional copper bit or a good smoking heat for electrical irons. Clean and tin the iron tip, then sizzle it in flux (a).

Tin the first wire's end. Lay one face of the iron — not the point — on the cable so maximum heat flows from one to the other (b). The flux will melt first. Leave the iron in place and touch the solder on the wire, not the iron. When the work is hot enough, solder will melt into it.

Don't transfer liquid solder to the cable with the iron, as the solder will pick up oxides.

9.7

9.7. Peripherals for soldering. Resin flux and cored solder (center) are primarily for electrical work. Acid flux (left) scours surfaces and will corrode if not washed off: it's generally used with plain solder (right and front) on plumbing and sheet work. The gas blowtorch (top) has replaced the blowlamp for heating traditional irons. Much smaller types carry butane lighter fuel in an internal reservoir: interchangeable bits make them very handy for soldering, for heat-shrinking tube and for cutting/fusing nylon rope.

If the solder won't melt reasonably quickly, just hold the iron where it is. Reaching melt temperature may take a while, particularly if the iron is only just big enough for the job. Then tin the second wire's end in the same way.

Lay the two tinned wire ends alongside each other. Don't twist them: if you do the pair will unwind under load, as a pull concentrates shearing forces in a small area, destroying the solder bond.

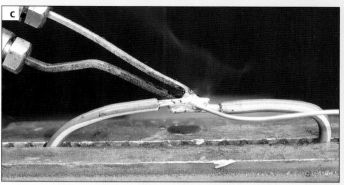

Bring a hot, flux-dipped iron down on the joint, holding it until all the solder turns liquid silver and runs in both halves. If necessary (and if you've got three hands) add more solder to the joint to fill gaps (c).

Take off the heat and **hold the wires steady** until the solder freezes, turning from bright silver to dull gray. Any movement during this step will give a crystalline "dry" joint with little strength and poor conductivity. Clean off flux residue.

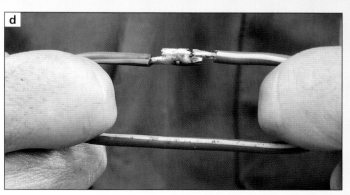

The end result takes a fair pull (d), where twisted wires would just unwind.

To insulate the joint, slide on some heat-shrink tubing before soldering (e). Buy it from an auto factor or an electronics shop. Gentle heating with a gas torch will shrink the wrap down over the bare wire, giving a finish much tidier and longer-lasting than insulating tape. If waterproofing is required, clean the area then wrap it in self-annealing tape. Also available from electronics shops, this bonds with itself to form a waterproof coat.

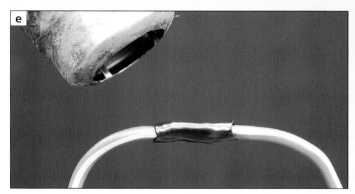

Adding a Cable Nipple

Soldering a nipple on to a Bowden cable is useful for a quick repair or when making controls for a one-off machine. This technique also covers attaching "bullet" wiring connectors, but leave out the last two steps. Use non-corrosive resin-based flux and flux-cored solder, as acid flux will attack and rot the cable strands.

Cut the cable end cleanly. An efficient (if rather unkind) way of doing this is to hold the cable between the cutting jaws of pliers or nippers, lay the pliers on an anvil or vice and give the jaws a sharp whack with a copper-faced hammer. Watch out for flying bits — wear goggles.

Clean and tin ¼"–⅜" (6mm-10mm) of the cable end (a). Old cable will be oil-soaked, so degrease and abrade it or the solder definitely won't wet.

Slide the nipple down the cable until ⅛"–⁵⁄₃₂" (3mm-4mm) is left sticking out. Gently gripping the cable at the right height in a vice makes this easy. Hold a hot, fluxed iron on the nipple and cable until both are smoking, at which point add solder: this should flow between

them (b). If it won't, carry on heating or re-clean. Add solder until it pops out at the nipple base and the nipple's hollow head is part-filled. Let the solder solidify.

Working with a knife blade on the protruding cable, separate individual wire strands and bend them back to form something that looks like a chimney brush or tree (c). *This step is vital for strength:* once buried in solder, the splayed ends won't pull out under load.

Now bury the strands. Turn the cable over and reheat it so the nipple slides down, settling the splayed wire ends into the nipple cup (d).

Turn the cable right way up and add more solder until the nipple cup is filled and all splayed strands are buried (E). Take off the heat and hold everything steady until the solder sets.

Finish off by grinding back any projecting strand ends. If you've forgotten to slide the cable into its outer sheath or cut it to the wrong length, kick yourself and start again.

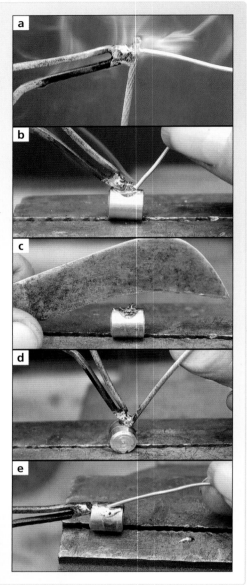

What's Going Wrong? Common Soldering Problems

Soldering snags always come down to a lack of heat, contamination or a shortage of flux. Or some combination of these. So problems aren't hard to fix.

Symptom: Solder won't melt on to the joint or melting stays local, next to the iron. Solder flows sluggishly or not at all,
Cause: Insufficient heat.
Cure: Check that the tip is clean, tinned and not covered in heat-transfer reducing oxides. Use flat face of iron for maximum contact area. Don't be impatient; heat transfer takes time. If none of the above work, the iron is either not hot enough or is too small for the job.

Symptom: The joint is up to temperature but solder won't flow, maybe spreading in some areas but refusing to wet others.
Cause: Surface contamination or inadequate fluxing.
Cure: Stop trying. Clean the target until it shines. Coat it with flux and try again with a clean, hot, fluxed iron. The problem can also be in the solder itself — dull gray, oxide-covered stuff won't flow easily. Shine it with emery cloth and try again.

Section 10

Welding Plastics

What happens to damaged plastics on the farm — the split mudguards on tractors and materials handlers, the torn waterproof covers on the drill or spreader hoppers, the leaking tank where someone got a little out of shape with the muck fork? Chances are the damage is either left, botched up with glue, or patched with a bit of sheet and rivets.

But there's a much better way. For a modest investment in kit you can make permanent repairs that are pretty much as strong as the original. And the very good news is that if you know how to weld steel the technique's easy to pick up, and it's not much harder if you don't.

The Swiss make the best kit in the shape of Leister equipment, so we'll feature this. While some Leister machines work their magic by ultrasound or laser, most

heat the work with nothing more than a directed blast of hot air. Much in the way of politicians.

For farm work we're interested in hand-held models. Designed originally for the manual repair of vehicle bumpers and other high-cost bits, these are like a rather clever hair drier — see (10.1) — with accurately variable output temperature and interchangeable nozzles.

Can I Join It?

Only **thermoforms** can be welded. Thermoforms soften on heating then harden as they cool, and will put up with many heat/form/cool cycles.

The other main group — **thermosetting** plastics — can be melted only once and harden as heat is added, generally in a mold. Thermosetting plastics crack if reheated. Though not weldable, they can often be repaired with the right adhesive, joint design and (ideally) some reinforcement.

Luckily, most of the moldings found on farms and in vehicles are thermoforms. To weld these successfully you must have a matching filler material, and use a temperature which allows fusion but stops short of burning. That's where a purpose-made welding gun comes in — a soldering iron or the wife's hair drier just won't do.

Choices

If you're fired with desire to join plastics you must have the right equipment. All the following are made by Leister. The Triac 1G3 gun is industrial tackle at a premium price — a complete kit with four nozzles, a rotary burr, case, stand and a selection of filler rods cost around $575 at the time of writing. If you prefer a pistol-grip model, the Ghibli gives similar performance at the same price.

If that sort of money is too steep, consider the EcoTool Weldy, a pistol-style hobbyist version which handles the same range of work as the Triac but delivers roughly half the welding speed. Pre-set temperatures and airflows for different materials are called up on a neat LCD display, making life simple. It'll also solder, shrink-wrap, weld bitumen roofing felt, strip paint and heat seized parts. A Weldy kit with ⁵⁄₃₂" (5mm) pendulum welding and tacker nozzles is $300: extra nozzles, a silicon roller and a bundle of mixed filler rods adds around $115.

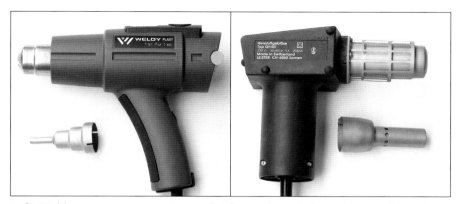

Left: Weldy programs temperature and airflow automatically to the material, with manual setting also possible. At $320 it's cost-effective for on-farm repairs, though workrate is halved compared with more expensive Triac. Green plastic roll at rear acts as a stand. *Right:* Ghibli pulls all Triac features into a pistol-grip gun, and as a bonus can take a de-horning nozzle (above). Kit costs $570.

Welding plastic is a fusion process, like stick or MIG in steel. The joint is heated to melt point, then if necessary a filler rod is fused into the parent material to fill and reinforce it. Unlike in an arc method, though, the filler does not need to be completely melted; fusion happens only where the rod and parent plastic meet.

We'll look first at the techniques. As with steel it's vital to pick a filler rod which matches the material to be welded: for more on that, see "Sherlock Time."

Heat for the Job

Welding plastics is no different in principle to welding steel. The same four things must be right: the filler rod, the rod angle, the speed of travel and the temperature. If the work isn't hot

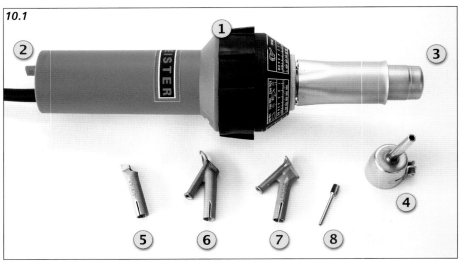

10.1. To weld plastics you'll need a welding gun (1) with adjustable heat setting (2) and heat outlet (3). Nozzle (4) slips over the outlet and is used for local heating and pendulum welding. Separate smaller nozzles push-fit over it. Part (5) is for tacking, (6) and (7) are for speed welding. A rotary burr for Vee-preparing joints (8) completes the set.

Sherlock Time

It's crucial (but not hard) to identify the material you're dealing with. Thermoforms fall into three broad groups — the polyethylenes, the polypropylenes and the polyvinyl chlorides or PVCs. If the part up for repair carries no identifying code (see below), see what happens when you try to bend it:

- PVCs are rigid and won't bend easily.

- Polypropylenes are flexible and show stress marks at the point of bend.

- Polyethylenes are flexible and show no stress marks when bent.

That'll get you into the ballpark. For confirmation make a short test weld somewhere out of sight, not behind the barn but on the inside of thing you're repairing. Run a series of short pendulum welds using different-material rods, leaving a stub on each. When cool give each a hard tug. Only the compatible filler(s) will have fused; the others will just lift away.

The most reliable identifier may be on the part itself. All modern vehicle

bits (and many other things) carry ID, primarily to help in recycling. Typically it's a group of letters, with perhaps a date stamp. Sample codes are in the table below. There's no absolute guarantee that you'll be able to weld an item once you've pinned down what it is — allegedly some car makers use additives to counter fusion repair, presumably to sell more spares. It's a wicked world. . . .

Don't know what material you have? Make short test welds with different fillers. The rod that won't pull off when cool is the one to use.

Table 10.1: Plastic Codes	
Code	**Material**
ABS	Acrylonitrile Butadiene Styrene*
ABS/PC	Polycarbonate alloy of ABS
PA	Polyamide (Nylon)
PBT	Polybutylene Terephthalate
PC	Polycarbonate
PE	Polyethylene
PP	Polypropylene
PP/EPDM	Polypropylene/Ethylenediene rubber
PUR	Polyurethane**
PVC	Polyvinyl Chloride
GRP/SMC	Glass fiber reinforced plastics***

Notes: *ABS is very commonly used for vehicle moldings **Not all PUR is weldable ***Not weldable

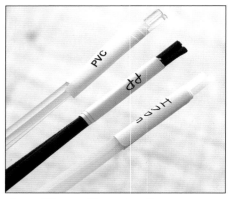

Three different filler rods will cover most farm jobs: mixed packs are ideal. Store them so the labels stay readable.

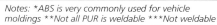

enough then there's little or no fusion; if it's too hot, the plastic distorts or burns. In most cases, smoke coming off the weld area says things are definitely too hot.

So what's right? Most plastics are welded between 518°F (270°C) and 752°F (400°C). You must have a proper heat source: a paint stripping gun simply hasn't the required temperature control and can't deliver enough hot air in the right place. Leister's equipment (used to make all the welds shown here) allows precise temperature setting, then maintains outlet air at the specified heat. Table 10.1 lists the requirements of different materials.

Table 10.2: Welding Temperatures

Code	Temperature °F (°C)
ABS	662 (350)
ABS/PC	662 (350)
PA	752 (400)
PBT	662 (350)
PC	662 (350)
PE hard (HDPE)	572 (300)
PE soft (LDPE)	518 (270)
PP	572 (300)
PP/EPDM	572 (300)
PUR	572/662 (300/350)
PVC hard	572 (300)
PVC soft	662 (350)
Xenoy (PC alloy)	662 (350)

Ways to Go

Three techniques — tack welding, pendulum welding and speed welding — cover the majority of work. Different nozzles are used for each.

Unlike arc or gas processes, welding plastic is a very physical job. You must put steady pressure on the filler rod (and sometimes the welding nozzle) to push it into the joint, where it fuses with the hot plastic below. You'll quickly get a feel for how much pressure to use with

different fillers and different techniques, and then it's a very satisfying thing to do. And as there's no arc to transfer metal across and no molten weld pool for gravity to pull on, the technique is the same for butt joints and T-fillets, for repairing cracks and working vertically or overhead.

There's one other big difference. Apart from when welding PVC sheet — the stuff used to make hopper covers, truck curtain sides and the like — there should be no fume. If a joint smokes, it's too hot.

Preparation

Fusion just ain't gonna happen where there's glue, paint or dirt. So get rid of them: work very gently with a 120-grit disc or flap wheel to clear an area around the joint.

Plastic welding doesn't have the penetration of an arc process, so repairs in all but the thinnest sheet must be opened out to create a bed for the filler rod to lie in. If you're fixing a crack, first drill the end(s) to stop it spreading (a ⅛" [3mm] hole is good) and carry on Vee-preparation beyond the end of the fault line.

Leister kits come with a rotary burr for crack preparation, or you can buy one

from a model shop. Mount the burr in a Dremel tool or similar, and spin it fast so it won't snag. Open a Vee-shaped groove along the repair line to about two-thirds the material's depth. Wear eye protection — plastic swarf is as nasty as metal.

Tack Welding

An effective tack in plastic fuses the bottom of the joint, seals it against leaks and holds the parts in accurate alignment. Tacking is always full-length along the joint, rather than in short stitches, and filler rod isn't used.

A tacking nozzle has a thin tip, backed by an exit slot for hot air (item 5 in picture 10.1). Once up to temperature for the specific plastic to be welded (see Heat for the job) the tip is drawn fairly briskly along the bottom of the joint, using very light down pressure (10.2). Angle the gun so only the tip's forward segment makes contact (concentrating heat where it's wanted) and position it so joint halves get an equal melt. Although it's important to align parts accurately before tacking, hot plastic's malleability often allows for limited re-jigging before the weld knits.

If the joint needs bracing up after tacking, turn it over and run one or

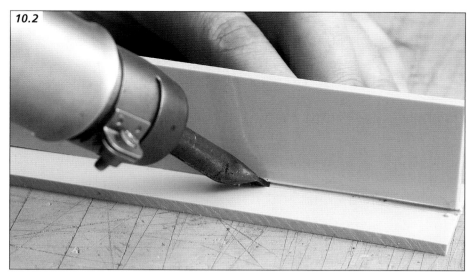

10.2. Tacking is always full-length. The heated nozzle is angled so only its tip touches, then is simply drawn along the joint with very light downward pressure. Hot air from a slot before the tip pre-heats the work.

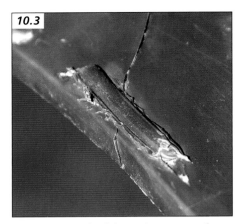

10.3. Support for joints edges comes from a fusion weld, stitched at one or more points on the reverse of the work.

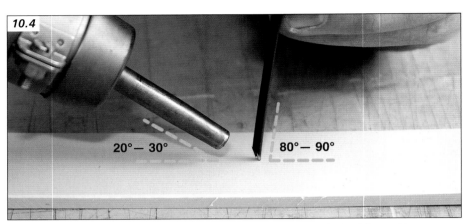

10.4. For pendulum welding, the gun stays at a low angle while the filler rod is almost vertical.

20° — 30°   80° — 90°

10.5. Pendulum welding in action: the arrow shows travel direction. See how the downward pressure needed to push the filler into the plate bows the rod?

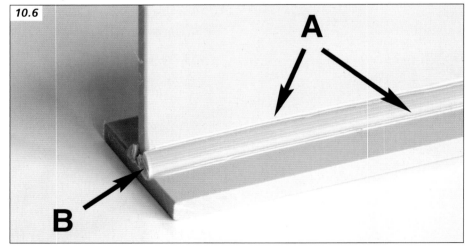

A

B

10.6. A good T-fillet weld made with a profiled filler. During work, adjust travel speed and downward pressure on the rod until molten ridges (A) are squeezed out where rod and joint meet. These ridges confirm good fusion and should appear in every weld. Excess rod was snipped off once it cooled (B).

more short welds at right-angles to the main line (10.3). See next for technique.

Fillers

Plastic welding suppliers offer rods in various materials and sections, either separately or as a mixed bundle. Round filler can be used in any joint; ⅛" (3mm) diameter covers most work. Triangular section fillers are ideal for either Vee—prepared, T-fillet or butt joints, the latter where a penetration gap has been left. The ⁷⁄₃₂" (5.7mm) variety leaves a smart profiled bead (see 10.6) and is best used through a speed welding nozzle.

Pendulum Welding

The most flexible of the three techniques and the one best able to get into tight corners, pendulum welding gets its odd name from the grandfather-clock swing of the nozzle during work.

Prepare and tack the joint as above. Choose a filler of the right material and with the best profile for the joint, then snip the business end to a tapered point: the taper lets it melt progressively into the start. Hover the rod at 80°-90° over the surface (10.4), then hold the nozzle air slot over the end of the joint until the surface just starts to sweat or slightly changes texture. To begin the weld, push the filler firmly into the joint. Use enough pressure to bow the rod in the direction of travel (10.5). If the heat is right, you'll see a ridge of shiny molten plastic squeezing up on both sides of the filler (10.6).

Move off down the joint. As you go, swing the nozzle slowly backwards and forwards along the repair line, preheating mainly the unwelded area but also washing the rod tip with hot air on every swing (10.7). Steadily push filler into the joint and hold travel speed steady.

If you run out of filler before the finish, stop before you burn your fingers. Snip off the stub as close as possible to the joint. Clip the end of a fresh rod, pre-heat the restart area and off you go. Fuse the very end of the joint properly, take off the heat and leave the rod attached to the work until it cools enough for a clean snip.

What happens where there's a hole to fill? It may be possible to heat a damaged, folded section and tease it back into place before welding. Where a hole still has to be patched and you've identified the material needed, your local plastic supplier (check the internet) should be able to set you up with sheet stock. Alternatively rob a bit from a broken piece of the same material.

Although pendulum welding lets you operate in tight corners and gives the most flexibility over working angle, it isn't particularly fast. Where large repairs have to be made in rigid material, **speed welding** may be better.

More Haste, More Speed

Speed welding is the easiest, fastest but least flexible of the three welding techniques. The big difference is the use of a boot-shaped nozzle, slipped over the small tube nozzle used for pendulum work (10.8). The forward part of the boot is an open tube which takes the filler rod, while the rear opens into a slot which ducts hot air to pre-heat the weld zone. Different nozzles accept round-section or profiled (triangular-section) rods.

Speed welding technique is much the same for crack repair, butt welds or tee-fillets. Prepare the joint properly or the weld will fail — abrade away any paint, grease or dirt, then Vee-out a crack to half-depth using a small, high-speed rotary burr.

Then tack the joint on both sides down its full-length to fuse and seal the bottom edges. Don't skimp this step; sealing is vital when repairing tanks and suchlike.

You'll need to wait a couple of minutes for the nozzle to come up to full heat.

10.7. No fancy protective equipment is needed to weld plastic — just the right tools and a modest helping of skill.

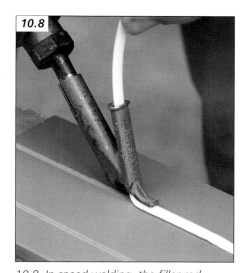

10.8. In speed welding, the filler rod (here white) is fed under firm downward pressure through a boot-shaped nozzle. The weld is carried right through the end of the joint so fusion is complete. The nozzle is slid up and off the filler, then excess rod is snipped off.

Meanwhile, snip the filler rod at an angle to produce a tapered, quick-melting tip. When the nozzle is ready, hover its air outlet over the start of the joint, angled so the bottom of the boot is parallel with the work. Position the gun so that both joint halves receive equal heat.

As soon as the plastic starts to change texture or sweat slightly, feed the filler into the nozzle. Don't do this too soon or it'll melt and stick in the tube. If that happens, stop. Clean out the goo while it's still warm and start again.

Push the filler rod firmly into the hot joint. As you do so, lift the nozzle a little so it rides over the just-fused rod. Draw the nozzle along the joint, keeping a steady, firm downward pressure on the filler. If you're not sure how much downforce to use, try pressing on the bathroom scales until they register around 1.1 lbs. (1kg).

It may help to put a little light pressure on the nozzle too — experiment to see what gives the best fusion. As with pendulum welding you're looking for a quick, even melt of filler, shown by molten ridges squeezing out on either side of the just-fused rod — check picture 10.6.

Draw the nozzle along the joint in the way you'd weld with a stick electrode, that is, trailing operation. That's from left to right for a right-hander, or right to left for a left-hander. Either way, a blast of hot air from the slot pre-heats the joint plastic for welding.

Weld right past the finish of the joint to make sure the area is completely fused. Then slide the nozzle quickly off the remaining rod and, once this has cooled, snip the surplus flush. Any remainder can be sanded off when cold.

STOPS AND STARTS

What's the deal when you run out of filler rod part-way down a joint? The principle's simple; slice out a notch in the warm, just-finished weld, then snip a matching point on the new length of filler. Pre-heat the restart area

and fuse the new rod into the notch before continuing down the joint. Figs 10.9-10.11 show how it's done.

FLEXIBLE SHEET REPAIR

So far we've looked at ways to join or repair rigid- or semi-rigid plastics. Also very handy on farms is the ability to fix holes and rips in flexible stuff like hopper covers, trailer sheet and lorry curtain sides.

These are usually made from PVC. Filler rod isn't used: instead the clean sheet and patch are heated until they melt, then the joint is made by quickly applying pressure to fuse them together.

You'll need a 1½" (40mm) wide slot nozzle, a silicon roller (necessary to stand 572°F-662°F [300°C-350°C] without wilting) and either a good flat bench or a stiff backing board, the latter good for jobs done away from

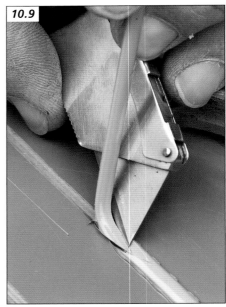

10.9. First step in making a restart is to notch out the warm, just-laid down bead into a shallow Vee, paring away the stub of old rod. A craft knife is ideal for this.

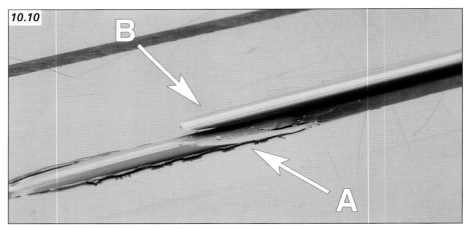

10.10. Here's the Veed-out bead (A), with the trimmed end of a new filler rod poised over it (B). See how the trim angles match.

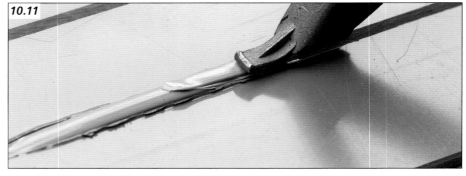

10.11. For the restart, the notched-out section of bead is heated then the new rod is fused into it. Any lumpiness can be smoothed away by re-heating the join and running a finger over the area, or by sanding when the joint is cold.

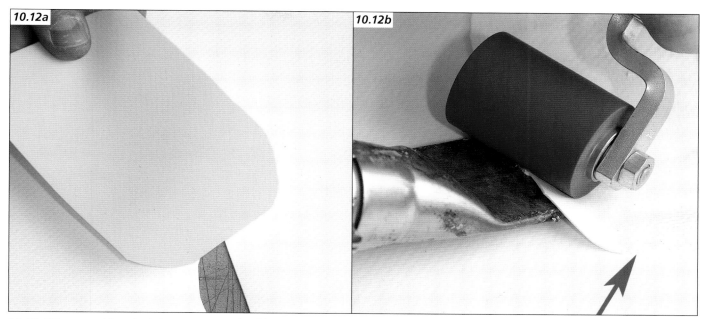

10.12. *When repairing vinyl sheet, cut a patch with a generous overlap. Use a ²⁵⁄₃₂" or 1½" (20mm or 40mm) slot nozzle and keep it moving. Heat equally between the patch and torn material until smoke just starts to billow out, showing that the joint halves are melting. Pressure the patch into place by a hard rub with a silicon roller. Fusion should show all round the patch edges. If not, re-heat the area from above and roll again.*

the workshop. Cut and clean a patch from similar material, sizing it to allow at least 2" (50mm) overlap round the repair site. Clean the parent sheet and position the patch over it.

Starting at one corner, slip the nozzle between patch and sheet (10.12). Keeping it moving from side to side, heat a small area until blue-white smoke just starts to blow from between the sheets — this is the one and only time that smoke is a good sign. Pull out the nozzle then quickly and firmly roll the patch down. A thin line of molten plastic should appear at the joint edge. If it doesn't, heat the area and roll again. It's vital to keep the nozzle on the move to avoid overheating. When the first area is done, work across the patch until the repair is finished.

PROBLEM SOLVING

Plastic welding suffers from the ills of all fusion processes — poor preparation, not enough heat or too much of it. Figures 10.13 and 10.14 show three common faults.

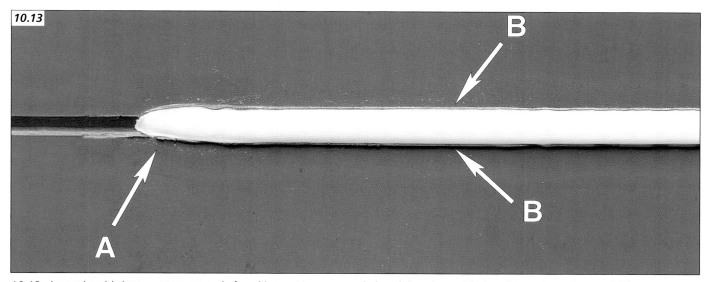

10.13. *A good weld shows a start properly fused into a Vee-prepared plate (A) and equal fusion ridges along the bead (B). Temperature, pressure on the rod and speed of travel were right.*

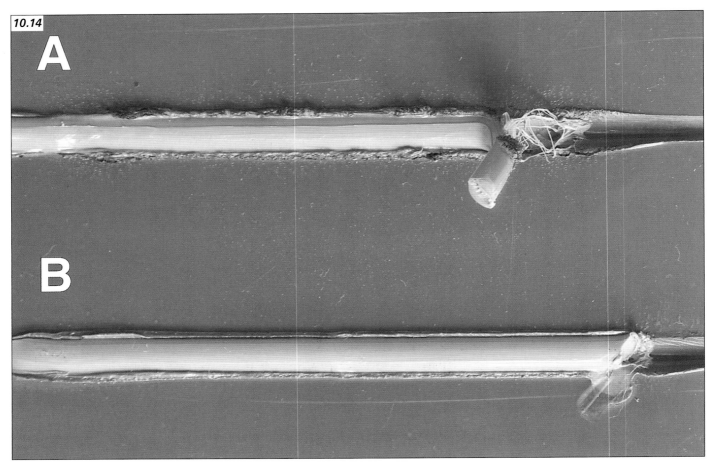

10.14. Too much heat produces scorching and/or bubbling along the bead flanks (A). Air coming from the gun is too hot for the material. In (B) the temperature is fine but excessive downward pressure has forced the filler to sink, leaving a recessed, weakened bead.

This is a car bumper repair, but it could equally be a tractor mudguard or a water tank. The bead can be sanded flat when cool.

Workshop Techniques

1. Taps and Dies

A good set of taps and dies definitely saves heartache. Their uses go way beyond the obvious one of cutting your own threads — for example, crossed-up nuts and bolts can be recovered, holes can be revitalized after broken studs are extracted, threads can be extended beyond a worn section, a damaged bolt can be drilled out and a new, oversize thread made; the possibilities are many.

Cutting a good thread isn't rocket science. Follow a few rules, work carefully and the job will be good. But rush at it and you'll probably make out-of-true, poorly cut and perhaps shallow threads that won't have much strength. And if you accidentally break off a tap in something that matters, extracting it will probably reintroduce heartache. So here are a few tips, starting with the oily stuff.

LUBRICANTS

Lubricating a tap or die in work cuts wear and greatly reduces the chance of it jamming or producing torn threads. All common metals except cast iron and brass benefit, as will your biceps and temper. Purpose-made liquids, soluble cutting fluid or greases are the best bet (11.1) but in a pinch you can use ordinary oil. Greases help by collecting some swarf. Cast iron and brass can be threaded dry, aluminum prefers paraffin to prevent tearing.

TAPPING

Whether you drill a fresh hole before tapping or use an existing one, its size is critical. Too big and the threads will be shallow and weak, too small and the tap will jam.

So how do you find the tapping hole to match a given tap? For metric threads it's simple: just subtract the pitch from the bolt diameter. So for a 10mm bolt

Identity Crisis

Before you can tap or repair a thread you'll need to which it is. The informal way is to offer up various taps to a pattern bolt until you find the one which meshes exactly with the bolt threads; measure the bolt diameter to check the required size and off you go. If it's info on a nut or hole you need, find a bolt which screws into it and measure that.

A more formal method is to use a thread gauge. These often come with a tap and die set, or can be bought separately. Like a feeler gauge, the thread version has fold-out fingers — only each is cut with a named thread profile. Find the one which interlocks snugly with your sample bolt (a).

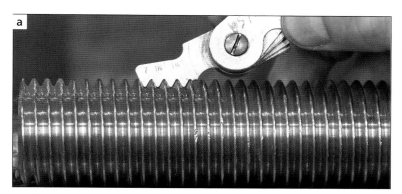

a. Thread gauges have different profiles. The one which exactly matches the unknown thread identifies it.

The third option needs an accurate ruler, good eyesight and a little patience. If you suspect the thread is Imperial (maybe from the age of the machine or bolt diameter), count the number of thread crests in one inch to find the tpi (b). If you suspect it's metric, measure the distance between two crests to find the pitch. Check the bolt's diameter on an unthreaded section, look this up in a data table then run through the thread forms until you find the one whose tpi or pitch corresponds to your sample.

b. An engineer's steel rule is best for counting threads per inch or measuring pitch — the distance between two adjacent crests.

11.1

The Naming of Parts

- **Taps** cut internal threads in holes.

- **Dies** cut external threads on round stock.

- Taps are held in a twin-handled **wrench**, dies in a **stock.**

- A **die nut** — a die with hexagonal spanner flats — is ideal for quickly clearing clogged or damaged threads.

- The **pitch** of a thread is the distance between adjacent peaks (or crests) and shows how far the fastener advances in one full turn. It's used to identify Metric threads.

- The number of **threads per inch** (tpi) is used to identify Imperial threads.

- A **pitch gauge** identifies unknown thread forms.

- Engineer's **data tables** are used to identify threads and find tapping hole diameters. Good suppliers have them.

- A **tapping drill** bores a hole of the right size to accept a thread.

11.1. Ideally use a specific tapping lubricant (right) or undiluted cutting fluid (left). Ordinary oil or grease will do in a pinch.

with 1.5mm pitch, the ideal tapping drill size is 10mm-1. 5mm = 8.5mm. Pitch will be marked on the tap, usually next to the item's size in mm.

Tapping drill sizes for Imperial threads are best looked up in engineering tables (11.2).

Whether you use the recommended drill or one a touch larger depends on the circumstances and material. A sharp bit in a drill press will cut to the expected diameter. A blunter one used manually is more likely to leave an oversized hole. Usually a slightly oversize tapping hole helps, particularly in hard stuff. The thread won't be made to full depth but strength need not be greatly compromised, and the job will be easier. The deeper the hole and the longer the bolt, the more threads there are to share the load so the shallower they can be. Suck it and see, cautiously re-drilling

11.2

I.S.O. METRIC COARSE THREAD FORM

r = Basic Radius = ·1443 p
hn = Basic Height of Internal Thread & Depth of Thread Engagement = ·54127 p
hs = Basic Height of External Thread = ·61344 P
p = Pitch

nom. dia.	pitch	basic major diameter	basic effective diameter	basic minor diameter of external threads	basic minor diameter of internal threads	recom-mended tapping drill size	clearance drill size
mm	mm	mm	mm	mm	mm	mm	mm
1	0.25	1.000	0.838	0.693	0.729	0.75	1.05
1.1	0.25	1.100	0.938	0.793	0.829	0.85	1.15
1.2	0.25	1.200	1.038	0.893	0.929	0.95	1.25
1.4	0.30	1.400	1.205	1.032	1.075	1.10	1.45
1.6	0.35	1.600	1.373	1.170	1.221	1.25	1.65
1.8	0.35	1.800	1.573	1.370	1.421	1.45	1.85
2	0.40	2.000	1.740	1.509	1.567	1.60	2.05
2.2	0.45	2.200	1.908	1.648	1.713	1.75	2.25
2.5	0.45	2.500	2.208	1.948	2.013	2.05	2.60
3	0.50	3.000	2.675	2.387	2.459	2.50	3.10
3.5	0.60	3.500	3.110	2.764	2.850	2.90	3.60
4	0.70	4.000	3.545	3.141	3.242	3.30	4.10
4.5	0.75	4.500	4.013	3.580	3.688	3.70	4.60
5	0.80	5.000	4.480	4.019	4.134	4.20	5.10
6	1.00	6.000	5.350	4.773	4.917	5.00	6.10

ZEUS PRECISION

DATA CHARTS AND REFERENCE TABLES FOR DRAWING OFFICE TOOLROOM & WORKSHOP

METRIC REVISION

11.2. Once you know the thread's size and form (yellow, left) look up the required tapping hole diameter (red, right). Data tables (inset) come from engineering suppliers. (TABLE © DORMER TOOLS.)

11.3. A taper tap (1) has many chamfered threads for a progressive start. A second (3) has fewer, while a plug tap (3) has one or two. If you have only types (2) or (3), you'll need to countersink the hole and take particular care to start the thread square.

a tapping hole larger if a sharp, well-lubricated tap won't cut freely.

THE LUXURY OF CHOICE

Taps come in many forms. For hand work we're concerned with only three—the no 1 or taper, the second and the bottoming (or plug) tap (11.3). A decent combination set will have all three.

Taps are used in sequence to start the thread then bring it to final size. The harder the material the more important this is, as the risk of breaking a tap goes up. But many sets only have a second and plug (the latter vital to tap to the bottom of a blind hole), while really cheap versions compromise with just a second. If you're stuck with only one form of a carbon steel tap, bore the tapping hole a little oversize and countersink it to help the tap start square, then proceed with caution.

BREAKAGE? DON'T PANIC (TOO MUCH)

Taps usually break in direct response to gorilla tactics. These may be prompted by misalignment, tap bluntness, lack of lubricant or reaching the bottom of a blind hole. When the worst happens there are still a few options: assuming

there's no stub to unscrew and that the broken section is reachable, try tapping the tap's flutes round with a pointed drift. If that won't work it may be possible to break up the brittle tap where it sits, though at the risk of damaging the surrounding item. If you go for that option it's essential to wear eye protection.

A third possibility is to heat the tap to soften it, then drill it out. Again, that depends on the situation. Or use a specialist tap extractor — which slides prongs down between the flutes to allow unscrewing) — though these can only operate where there is room. The best route of all is not to break the tap. Procedures are covered in pictures 11.4-11.6.

USING DIES

Now for the other half of the story — cutting external threads with a die. First though, see "Pick Wisely."

Identification of unknown thread forms is the same for taps and dies. Here's a quick recap — pitch gauges are the quickest and most certain way to put a name to what you have, followed by counting the number of threads per inch of fastener and looking up the result in engineer's tables. Cues can come from the machine's age and origin; older equipment from the UK or USA will probably use Imperial fasteners (Whitworth, UNF, UNC or BSF; BA for small items, BSP for pipe), while modern kit or older stuff from Europe will probably use metric coarse or fine thread forms.

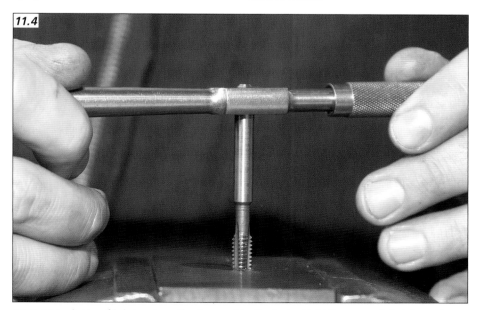

11.4. Use plenty of lubricant on the tap and in the hole. Hold the wrench lightly and steer, not force, it round while keeping it level. If the tap is sharp and the hole size is right, it'll bite quickly.

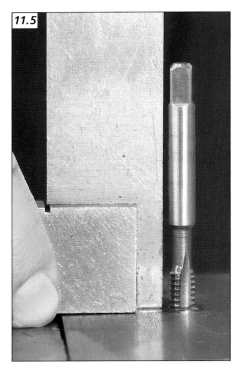

11.5. Once the cut is established, take off the wrench and check the tap for squareness in two planes. If it isn't running true, gently center it when you restart. Check at every turn and correct as you go until it's cutting straight. Don't leave an error too long!

As with taps, lubrication is a vital part of using dies. A good helping of it minimizes wear on the die and reduces thread-tearing. Guidelines here are unchanged; purpose-made fluids or greases are ideal for steels (though ordinary oil is better than nothing); aluminum is partial to a drop of paraffin, and brass and cast iron can be threaded dry.

Pictures 11.7-11.12 detail the way to cut an external thread on round bar or pipe. Keeping the die square with the work is harder than with a tap, as less of the die is chamfered — but is no less vital. Skill and care produce a good result.

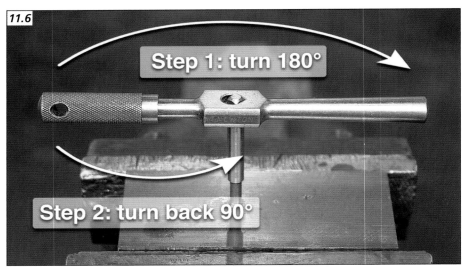

Step 1: turn 180°

Step 2: turn back 90°

11.6. Good technique keeps the tap cutting freely. Make no more than half at turn clockwise every time (step 1, above) then reverse direction until you feel the new-cut chips break (step 2 above). Don't force the tap: if resistance grows, back off, re-lube and try small cuts. Clear chips from the bottom of a blind hole or the tap will jam. Once the tap exits or reaches the bottom of a hole, back it right off. Re-lubricate, run it in and out a couple of times and you're done. Finally, clean the tap ready for next time.

11.7. The die must go into the holder (or stock) the right way round, or starting the cut will be difficult. The stock is partly closed at the top and open at the bottom. One side of the die has its first few threads chamfered to help it start cutting; this chamfer is usually on the side with lettering (1). Clean any swarf from the stock and die so they come together squarely, then fit the die chamfer-side towards the work with its split lined up with the stock's center screw (2).

11.8. Three screws locate the die and vary its diameter, opening it to ease the first cut then closing it to bring the new thread to size. For the first cut, tighten the center screw to widen the die and just settle the outer screws into contact.

11.9. Starting the cut will be much easier if the bar is chamfered generously, as here. Die and the bar/pipe to be threaded must have the same nominal diameter. That is, a ⅜" (10mm) die will only cut easily and produce full-depth threads on ⅜" (10mm) rod, and so on. Clean off any black mill scale.

11.10. Lubricate the target — here neat cutting fluid is used.

11.11. To start cutting put light downward pressure on the stock and steer it round, keeping even pressure on the handles. It will bite quickly only if it's sharp and fitted the right way up in the stock, and if the bar or pipe has been chamfered evenly. It's crucial to check that the stock stays at right angles to the target, both initially then all though the cut. If it tips, the thread will be skewed. Check for squareness every turn, correcting as you go. Once the thread is running true don't push down on the stock; let it find its own way.

1: Half-turn (180°) clockwise

2: Turn back 90° to break chip

11.12. Cutting action is the same as with taps. For right-hand threads make half a turn clockwise, then a quarter-turn back to break the swarf chips. Keep the die and rod lubricated. Once the first cut is done, unwind the stock. Back off the center screw and wind in the outer screws evenly to close the die a little. Re-lubricate and re-cut. Depending on die sharpness and material toughness, two or three cut/adjust cycles may be needed to form the final thread. Clean out the die: job done.

Cheat a Little

A lathe can help cut a parallel thread. Chamfer the item to be threaded and center it in the chuck. Position the die stock over the target, then extend the lathe's tailstock mandrel until it gently holds the stock square to the work. Turn the chuck by hand to start the thread, then keep the mandrel in very light contact with the stock as the cut progresses. The same technique can be used with a good drill press.

Oddments

Round split dies are fine for diameters up to ½" (12mm). Bigger threads are best formed with two-piece dies held in a rectangular, adjustable stock; this give more scope for progressively bring the thread to size.

Die nuts give fast thread clean-up.

Cutting fine threads needs particular care. If you have many to do it's worth investing in a stock which takes replaceable guide bushes in its lower face: the bush fits over the item to be threaded, holding the two square and preventing crooked threads.

Die nuts (picture) have hex faces to take a spanner and are not split. Use them only for cleaning up existing cruddy or damaged threads.

Recycle or Throw Away?

A tap or die blunts when its first few cutting edges chip or round off. You'll know when it's on the way out as starting a thread is harder, the thing takes more effort to turn and will be more likely to tear new-formed threads.

Trying to re-sharpen blunt taps manually will only make things worst. The best bet is to throw smaller ones away, but send bigger, expensive ones off to a specialist for re-grinding.

Pick Wisely

Taps and dies come in different qualities. The key choice is between carbon steel and high speed steel. High speed steel (HSS) is the engineer's favorite; items made from this very durable material are usually ground to size, so can cut an accurate thread and will hold a edge for longer.

Taps and dies made in high carbon steel (HCS) cost substantially less, produce a less accurate result and blunt sooner. But don't dismiss them — away from a production environment, carbon steel items are fine when used carefully. If you aspire to a HSS set but can't afford one, a good compromise is to buy a big-name carbon steel set and replace individuals with HSS versions as the originals blunt.

Your set should cover today's common thread forms, that is, metric coarse/fine and UNC/UNF. A set handling older forms like BSF and Whitworth will be useful if you have dated machinery or are into vintage kit. Individual taps or dies can be bought for one-off jobs, including cutting left-hand threads. When choosing dies go for split versions rather than solid ones; the latter are best used for cleaning and re-cutting threads rather than forming them from scratch.

And sizes? Farming's needs are pretty much covered by metric sizes 6 to 24 and imperial ¼" to ¾". Not many suppliers offer this range in one box, whereas sets covering metric 6-12 and imperial ¼"-½" are very common. So an option is to start with one of these, then add bigger individual taps or dies as needed.

For last-a-lifetime quality, shop at an engineering tool supplier handling premium brands like Dormer, Presto or Apex and have items reground as they lose their edge. Expect to pay upwards of $900 for a big set, or around $100 for a single threadform in limited sizes. Modest pockets are better served by mass-market suppliers, who offer coverage in the $100-$150 bracket. Either way, avoid cheap bargains; in the long run they are unlikely to be cheap.

2. Drill Sharpening

Trying to make a hole with a blunt drill is only for the foolish. At best, progress is slow and the blessed thing runs off-center. At worst your eyes bulge, muscles knot, the drill squeals then either breaks or expires in a whisp of smoke. Compare that to the bliss of using a new one . . . the way those twin spirals of swarf wriggle upward, the soft hiss as hardened steel pares away steel, the satisfying speed of cut, the simple smoothness and truth of the finished hole.

What does it take to get from a blunt drill to a sharp one? Not a lot — a bench grinder with the right wheel(s), and some practice. Various jigs and rigs can be bought to make the job largely foolproof, but knowing how to sharpen freehand is very useful. So that's what we're looking at here.

Half the battle is knowing what you're trying to achieve, then fixing that picture in the mind's eye and matching your handiwork to it (11.18). But first the grinding tackle must be put into shape.

GOOD AND TRUE

Which grinding wheel should you to use? Good-quality twist drills are made from high-speed steel (HSS) and are best sharpened on a 60-grit aluminum oxide wheel. If you're looking to buy one of these, a typical type designation is A60-L5-V30; engineering suppliers are the best source.

The wheel MUST be in good physical shape, both to do the required job and to keep the user in one piece. Shrapnel from a burst grinding wheel takes no prisoners, so a whole slew of regulations governs wheel mounting and use — see Section 2, Abrasives. Sensible preliminaries before dressing (refacing) a wheel include testing it for cracks (11.14) and making sure that cushioning paper washers sit between the wheel and clamps. For proper procedures, see OSHA 1910.215A, "Safety in Grinding Wheels."

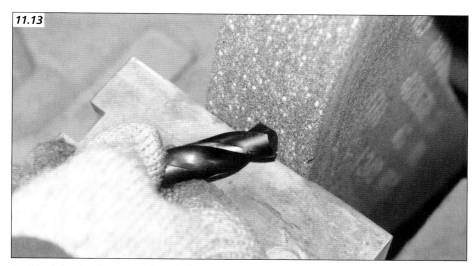

11.13. A fresh edge on a blunt drill speeds work.

The grinding surface must be flat to generate straight cutting edges on the drill. It follows that a wheel face resembling a map of the Norwegian fjords is not much use, so you'll need a dressing stone or a star wheel to level the surface and get rid of dulled abrasive grains (11.15, 11.16). Costs are around $6.00 and $19.00, respectively, from engineering suppliers.

Once the wheel is dressed, adjust the grinder's tool rest so the stone has minimal free running clearance (11.17). Keep the gap small to lower the chance of the drill being pulled into the space between stone and rest — something which will likely damage it, you and the wheel. With the bench unit's own eye shield in place or safety specs/goggles installed on the user, sharpening can begin.

THE CUTTING EDGE

How does a drill cut? By its two lips peeling away material, much as a wood chisel produces shavings. Although a small army of things influence how well any drill does this, three factors stand out — the clearance behind each lip, the total point angle, and the balance of lip length and individual point angles. A fourth factor (web thickness) starts to play a part as the drill wears down.

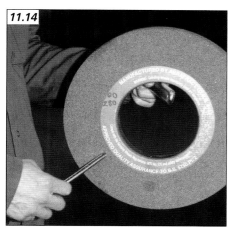

11.14. To check a grinding wheel for cracks, support it on an edge and tap gently with a length of hardwood. If it rings it's OK. A dull clunk suggests trouble — so do not use the wheel.

11.15. Two options for wheel dressing. The star wheel (left) is used on the tool rest. The cheaper stone (right) has to be harder than the wheel it's used on.

The best way to see how each of the key areas should look is to eye up a new drill, using 11.18 and 11.19 as guides. The bigger sample drill you use, the easier the detail is to see. With an automatic drill sharpener the right angles are relatively easy to achieve. Handwork is more hit-and-miss (particularly with drills under ⅛" [3mm]), so here's some suggestions to help.

Clearance. Any cutting edge needs a clearance behind it, otherwise it just rubs over the surface rather than slicing in. For the purpose of hand sharpening, a twist drill's clearance is the gentle fall away from the cutting edge and back across the flank (11.18). If this clearance angle is too small the drill just skates. If it's too large, the cutting lip has little support and will chip or break. For general work in mild steel, aim for a 10°-12° slope.

Point angle. Looked at from the side, the cutting lips steeple upward into a roof (11.19). If the angle they make is too steep, the lip edges turn concave seen from above and won't cut efficiently. If the point angle is too shallow the edges go convex, so cutting efficiency again takes a hit and the outer lip corners are weakened. For drilling mild steel the ideal angle is 118°, though this is not a fixed quantity. Shallowing it helps produce clean holes in thin sheet and will ease the drill through hard stuff.

Three DIY gadgets help get point angle right, though you'll soon do it by eye. One is a small template, easily knocked up and used to check individual angles during grinding (11.20). Number two is much the same, only made by soldering a pair of big nuts together and using the 120° angle between the flats as the guide. Number three is a 60° line scribed on the grinder's tool rest, along which you lay the drill while grinding.

Lip length and individual angle. Do your holes run off-center and end up oversize? Does most of the swarf come from one drill flute? If so, one cutting lip is longer than the other, one lip angle is steeper than the other, or both. Either

11.16. To true-up the wheel face, work the dressing stone gently across it.

11.17. After dressing, adjust the tool rest to give minimum running clearance. This lowers the chance of accident through work being trapped between the wheel and rest.

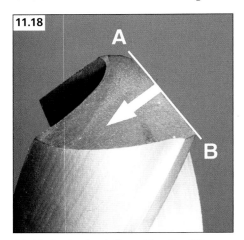
11.18. Each of a drill's cutting lips (A) must have a clearance behind it. This is the gentle slope backwards from the lip edge — see how point A is higher than point B? The right manual grinding technique automatically produces good clearance. Aim for a 10°-12° backslope.

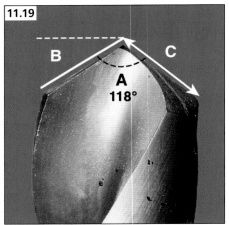
11.19. The tip angle under the roof (A) has to be right if the drill is to cut efficiently, though it can float a little around the 118° ideal and still give good results. When freehand sharpening it's very easy to get individual lip angles (B) or lip lengths (C) wrong. Then the drill will vibrate, run off-center and bore a more-or-less oval hole.

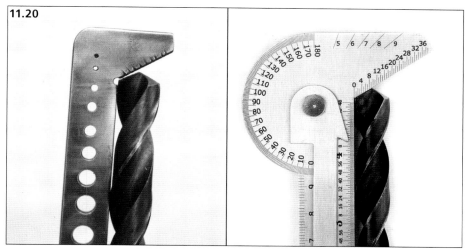

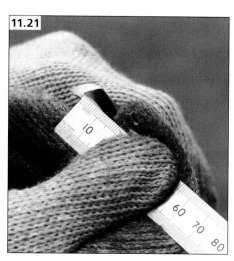

11.20. Made with care, a simple guide gives a quick, accurate check on individual cutting lip angles (left). Or you can buy one, complete with ruler for measuring lip length (right).

11.21. Use a steel rule or callipers to monitor lip length during grinding. (Fashion note: cotton gloves protect against oils and sharp edges).

way one cutting edge (the one with the shallowest angle) is doing most of the work. For the lips to cut equally, they must be physical twins. Angles can be checked with a template as above, the lip lengths with a steel rule or calipers (11.21). The smaller the drill, the harder this is to do.

THE NOT-SO-LONG GRIND

Take a close look at the drill to be sharpened, using a new one for comparison. All damage to cutting lips and corners must be ground away. If this will take a lot of work (as it will if the drill had sheared) rough-grind it first then fine-tune.

The grinding operation is in two parts:

• Offering up the drill at the proper angle

• Making the grinding stroke

If the stroke happens as it should, good clearance is produced automatically.

Angle first. Support the drill with one hand, hold the shank end with the other. Plonk the support hand on the grinder's tool rest, angling the drill so it sits just above horizontal with one cutting lip parallel to the wheel face (11.22). If you've scribed a 60° guide line on the tool rest, use it now.

The **grinding stroke** is marginally harder to describe than to do. Pictures

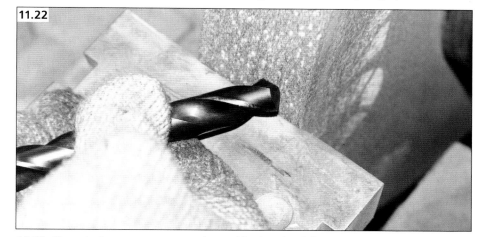

11.22. The grinding sequence starts by positioning the drill so one cutting lip is parallel with the wheel face. A 60° scribed line shows the right approach angle — particularly useful if a drill has been mis-ground or broken.

11.23. The right hand steadies the tip, the left guides the shank. Starting with the shank slightly below horizontal, begin a smooth, continuous upward movement. Lift the tip and feed the cutting lip gently into the wheel.

11.23, 11.24 and 11.25 break it into stages, though the reality is a smooth single stroke. Bring the cutting lip into light contact with the wheel (11.23) Then, while keeping stone and lip in contact, lift the drill tip and lower the shank. This scooping action moves the tip forward and upwards in an arc (11.24 and 11.25). Think of someone standing inside a huge pipe, brushing upwards from waist height with a broom; scale the vision down so the broom is drill-sized, and you've got the action. During the stroke, keep the approach angle constant and don't rotate the shank.

Make several light cuts on each lip, checking angles and lip lengths as you go. Go easy on the pressure — the cutting edge must not get too hot. If it blushes blue, toughness has been lost and the edge won't last.

Which brings us to cooling. Some say that drills should either be ground dry or under a torrent of coolant, but never anything between. Others reckon it's perfectly OK to take out modest tip heat in a pot of water but never to quench an overheated drill. A little and often is the watchword if you take the second route. Should the drill gets too hot before cooling, the softened cutting edge will turn brittle and crack or flake in work.

THIN THAT WEB

It's easy to forget the drill's central strengthening core or web, which gets thicker from tip to shank. As the drill shortens from repeated sharpening, web thickness at the tip grows. Why should you worry? The web is a dead, non-cutting area which has to be forced though the work, so the thicker it gets, the harder the drill needs to be pushed. Where a drill is only to be used for enlarging holes then no problem; the web plays no part. But where you're boring to size in one operation, web thickness does matter.

Thinning is usually needed after drill length has fallen by about one-third. Manual thinning calls for a shaped

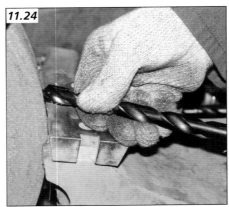

11.24. Here we're halfway through the stroke, with the tip scooping upwards across the grinding wheel face. The left hand maintains the original approach angle and does not twist the shank.

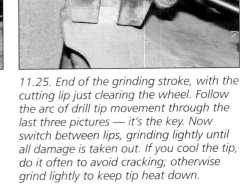

11.25. End of the grinding stroke, with the cutting lip just clearing the wheel. Follow the arc of drill tip movement through the last three pictures — it's the key. Now switch between lips, grinding lightly until all damage is taken out. If you cool the tip, do it often to avoid cracking; otherwise grind lightly to keep tip heat down.

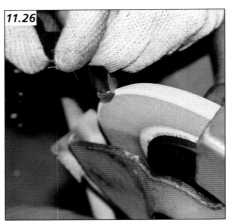

11.26. Web thinning calls for a wheel dressed to an angle of around 45°, with a radius on the high shoulder. Relieve the web equally on either side of the central chisel edge, grinding out a shallow arc.

11.27. A mid-range bench sharpener thinned this drill's web, splitting the point to form two small extra cutting lips (arrow). Now it will self-center and can be used without making a pilot hole first.

grinding wheel, which can be formed by careful work with a dressing block.

To slim a web, hold the drill vertical and just knick in along both sides of the point (11.26). Aim to produce a shallow arc, taking equal volumes from each side of the tip and ending up with a central chisel point measuring about one-ninth of the drill diameter. Alternatively, a well-designed sharpening machine will achieve the same thing for you by splitting the point — see picture 11.27.

Knowing what you're looking for when sharpening drills is the key, practice is the lock in which it fits. Given the right grinding wheel and a little patience, blunt-drill irritation will be gone.

Section 12

Basic Blacksmithing

Not so long ago the blacksmith kept a farm going. His farrier's skills kept horses at work, his forging ability shaped and repaired iron. All manner of things passed across the smith's anvil — chains, tools, bits of implements, household stuff, ironwork of all sorts. Bigger farms had their own smith; smaller outfits employed the village man.

Time passed, tires replaced hooves. Mass-production methods priced out handwork. The electric arc set gave every farmer the option to weld if he wanted, so gradually the forges grew cold. A lot of the contents went for scrap or ended up in pubs and gardens as decoration; but luckily many anvils moved sideways into farm workshops. That's excellent news for today's users, not least because the sound of a few blows (repeated at strategic intervals) has always been a good way of convincing a passing boss that great works are underway.

Village forges may have largely gone but the practical skills are alive as very well. Our tutor here is Frank Dean, a blacksmith with sons working in the Sussex smithy established by their grandfather at the turn of the century. And while the written word always comes a poor second to direct tuition from a skilled man, the pictures here will help. Try the techniques, using the illustrations as a target.

Equipment and Basics

If you're lucky enough to have a forge, then you'll appreciate how quickly and easily stuff can be brought to red heat in it. If not, all of the jobs shown here can be managed with the gas torch and a modicum of patience. It'll just take longer and cost a bit more in fuel. Or you can make a simple forge from a big lorry brake drum, a length of through-bored bar for the tue, a 12V car heater

blower and an old steel drum to house the blower and other bits.

If you're using gas, choose a cutting or (preferably) a heating nozzle that puts heat quickly into the job. A hearth made up from old firebricks helps to hold warmth round the work.

Getting comfortable is just as important as it is in welding. Arrange the anvil so its table is directly under your working hand. The pointed bick is then to the left of a right-handed man or vice versa. Strike blows from the elbow, rather than the wrist. Along with a proper grip this helps control the hammer head. Work in a steady rhythm — frantic activity only knackers you — and don't let the work get too far out of shape. Bring it back to close to what you're aiming for at the end of each heat.

Light taps on the anvil between strings of blows can keep a rhythm going. But when driving a punch, don't make a preliminary "sizing up" tap as this often jogs the point off-target.

Finally, keep the anvil clean and free from scale and the floor clear of tools and scrap. Tripping up with hot metal is at the very least painful.

Taking Heats

When is steel is at a good, easy-working temperature? The best way to find out is by experiment. Try working a piece of bar and see how its plasticity increases as surface color changes. Dull red is hard work. Blood red is better, and cherry red easier still. By yellow (around 2,372°F [1,300ºC]) the steel works very freely. But go one step too far and it becomes incandescent, with sparks flying off in all directions. A few degrees more (or seconds in the fire) and the metal will crumble away.

Make a Plan

Decide what you're going to do before heating. Think which part of the job needs to be tackled first, as it's often difficult or impossible to do one thing after something else. For instance, you can't forge a bar into an eye and then taper the inner tip: things have to happen the other way round. Also work out which part of the anvil you'll use, and which way to hold and present the work. Puzzling this out on the spot loses valuable working heat.

Pictures 12.1-12.6 run through the basics on the forge, anvil and tools. Then separate panels detail different jobs, starting with **drawing down** a

12.1. Air blast for the fire comes from the central tuyère or tue. To light the forge, first rake out any old fused clinker. Then excavate a pit to below tue depth. Check that the tue's exit hole is clear, and set a gentle draught on the blower to keep it that way.

Paper and dry kindling starts the fire. Fuel is ideally smithy 'breeze', but if your coal merchant has never heard of that or can't get it, Sunbright singles will do. Pile fuel into a banked mound. Lay extra around the edges for raking-in later. During work don't let the fire burn hollow: keep it banked up on three sides, leaving clear space on the working side. If the fire spreads out too far during work, damp down the edges.

12.2. The tue in a big forge is usually water-cooled, circulating coolant by thermal action. Check that the reservoir is topped up. You can make a portable forge from a truck brake drum and angle iron frame: the tue can be an uncooled 2" (50mm) bar with a ½" (12mm) hole bored through it for air passage. Draught can come from an old 12V motor with a simple butterfly valve to control blast.

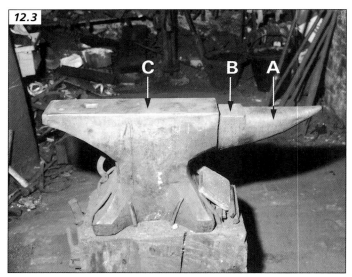

12.3. This 2½cwt London-pattern anvil sits on an elm post, and there's as much of that under the floor as above it. Working from right to left, the anvil has a tapered bick (or beak, or beck: depends where you hail from - A). Then comes the step (B). This is left soft during manufacture to provide an area where a cut-off chisel can be used without damaging itself or the anvil. The table (C) is the main flat working surface. It has two hardie holes: the round one primarily a clearway for punching operations, the square one to locate anvil tools securely. A farrier's anvil has no step.

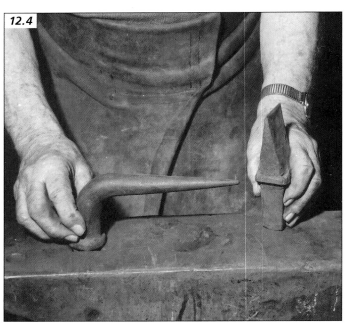

12.4. Commonly used tools for the hardie hole. A chisel-faced hardie (right) is used under hot metal when cutting, so that metal is sliced between it and the hammer face. The anvil stake (left) provides a small-scale bick for finer work. Or a twin-prong 'horn' tool (not shown) can be used for bending material.

12.5. Tongs are tools you can't manage without. They must match the metal they hold or it'll slip about, which is irritating and potentially dangerous. From left to right: bolt tongs with a flared section for the bolt head or any shouldered work; goose-billed tongs for holding flat rings and the like; small close-bit tongs for horseshoes or jobs where there's no great length of material to catch hold of; grooved hollow-bit tongs for round bar; and hollow bit tongs with a three-sided jaw for square material. Tongs can be suited to different-sized material by heating and bending their reins (handles). Secondhand tongs still turn up at farm sales or can be bought new, while experienced smiths make their own. In work, keep tongs cool.

12.6. Getting a grip. Frank Dean's hammer is balanced in his hand. His thumb is on the top side of the shaft, pointing toward the head, allowing good control and rapid rotation when the pein end is needed. "Holding the hammer too close to its head robs you of control", he says, "But holding it at the far end of the shaft is just as bad. Choose a shaft that's nicely oval, and lies comfortably in your hand. If the hammer is balanced there's no need to grip it tightly." Most hammering action must come from the elbow, not the wrist.

round bar into a point. This basic exercise is useful in its own right and a jumping-off point for all sorts of other operations.

General pointers cover all the work:

• Start with straight metal so you can concentrate on the task, not problems in the steel that were there before you began. If you must, true the steel first. As things get out of shape during work, stop and put them right.

• Think about exactly where your blows should land. If the plan is to stretch the steel, make a sandwich of it between the hammer and anvil to squeeze metal and change its section. But if the plan is to bend something, then place it and hit it so each blow has a bending (rather than flattening) effect. Hit "off" the anvil — pictures 12.11 and 12.12 make this clear.

• Two tips on marking out. To know how much metal is needed to make a ring of a given size, multiply its planned internal diameter by 3.14 and add the thickness of the material. If you're aiming for a given external size, subtract the material thickness from it and then multiply by 3.14. Where precision is not vital the anvil table can be used as a rough length guide. It's generally 5" (125mm) across.

• Often you'll need to mark metal to show where an operation (say, bending) has to begin. Center-punch the inside of the

Drawing down a point

12.7. Tapered points or chisel ends are easily made by drawing-down, which is the old way to give new life to harrow tines. The metal is brought to heat and held at an angle to the anvil. The steeper the angle, the steeper the point will be. See how the metal is held near the edge of the table, and the hammer face is angled to form the taper? This shapes the steel on two sides simultaneously. Between blows the work is constantly turned, always in the same direction (like a lathe) to draw the metal down evenly. Heavy stuff is worked first into a square, then an octagon, then a cylinder. Smaller metal can be tapered directly. For a long, gentle taper, first form the tip and then use the bick for the long work; its rounded profile draws down metal much more quickly than the table. "Worth a second hammer", Frank Dean says. If the tip splits, take a very good heat and draw it back together with light blows. With practice, multi-sided material can be reshaped to round or vice versa, using just a hammer and anvil.

12.8. The finished article. A coarse wire brush on the hot work shifts scale to leave a good finish.

bend. That way, the mark closes up as the metal goes over. If you punch-dot the outside, the dot elongates into a nasty knick — a likely start-point for failure.

Now for some specifics, starting with **edge bends.** Persuading a flat strip to bend against its width is not so hard (12.16-12.21). The method here produces a bend with a rounded outer corner. If you need a square corner, first thicken the metal locally at the bend point by "upsetting" or "jumping-up." Page 185 lays out how.

Before twist drills arrived, the blacksmith punched holes with a steel drift and muscle power (12.22-12.33). It's still a useful way to make square holes, rectangular holes, round holes . . . or big holes bigger. The old way may be slower than a drill but it's more flexible. Neither is it wasteful; a surprisingly small chip of steel pops out as a hole forms. Instead metal is displaced by the punch, leaving a void the shape and size of the thing that produced it.

Punches are easy to make. Shank length is a compromise — too long and the punch will bend, too short and you'll singe your hand. Slot punches are used first, do the hardest work and so need to be made from the best material; something combining strength with resilience. Frank Dean suggests an old file makes a good starting point. "Take out brittleness by bringing it up to a uniform blood-color heat, forge out any serrations, and then cool it very slowly under a bed of ashes or lime to prevent

Making Eyes

12.9. Step one in forming an eye is to true-up the metal, in this case ½" (12.5mm) bar. Notice how Frank Dean uses the hardie hole as a focus for straightening, lifting his left (or back) hand a little to give more clearance. Both give the metal somewhere to go. "It's hard to straighten something on a flat surface", he says. "Hitting down on the anvil's table will only achieve anything if it's worn into a saddle shape."

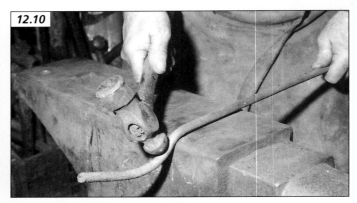

12.10. Unlike rings, eyes seldom need to be made to specific dimensions. Just make them so the eye looks right in proportion to the bar's thickness and length. Estimate the internal diameter, then work out how much metal you'll need from the formula on page 177. Then shaping begins. The first 5" (125mm) of the table after the step generally has rounded shoulders, so use this area to avoid putting a nick in the underside of the bar. With the metal at bright cherry red, strike the first blows as shown with the hammer pein. Hit just off the table's edge so the bar's free end kicks upward and the bend starts to form automatically. Don't hit too far off the table or a straight portion will be left just beyond the start point. Aim to make a gentle 'S'-shape.

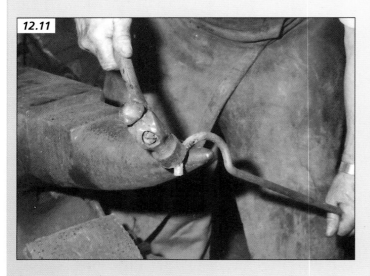

12.11. Take a fresh heat and move up to the bick. Here Frank is putting shape into the eye. See how his blow is directed to the side of the bick? This bends, rather than crushes the metal.

air-hardening," he suggests. "You can then forge or grind the tip to size."

Don't let the punch get too hot in work, he stresses. "Dip it in water every few blows, but don't quench it right back to cold or it'll turn brittle again."

A punch likes to stick in thicker material. A pinch of coke dust in the embryo hole helps it let go, and the black dust makes a visible marker through thin, hot metal when the work is turned over to finish operations. If it's round holes you need,

one or more mandrels (or drifts) will follow the slot punch. These enlarge rather than pierce, so can be made from mild steel. First to be used is a flat-ended mandrel, which takes the slot gently to a circle. Further enlargement is by tapered drift.

By swelling steel rather than chopping it away, **punching** allows much bigger holes to be made than with a drill. But as with all forge work, thinking ahead pays off. If it looks as though punching will leave too little metal round the hole,

steel thickness can be increased locally by **upsetting or jumping-up**.

Pictures 12.34 and 12.35 show how to do it. Upsetting is a powerful weapon in the smith's armory which finds plenty of uses. It's often the first job in a sequence, done to build a reservoir of metal for following operations. Where you can do it, upsetting can be a much better way to thicken a section than laying down weld metal.

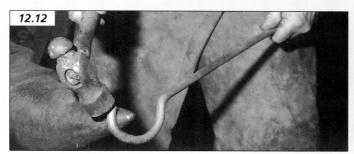

12.12. Now roll over the bar's tip. This is forethought in action, as it's impossible to do the job once the eye closes down. Again, the hammer strikes off the bick, rather than straight down on it. Frank's back hand is tipping the bar to provide clearance.

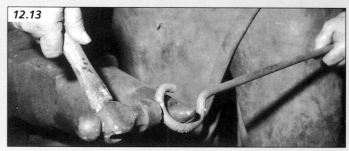

12.13. With the tip turned it's time to close the eye. Start by bringing it round the bick.

12.14. Then roll it shut by gentle blows to the top (a) and front (b), supporting the work on the table edge or bick as necessary to preserve the initial bend.

12.15. The finished article is a pleasure to look at. And beauty is more than skin deep. Centering the eye over the shank lets load fall in direct line with the shank: an off-center eye is more likely to pull open. The same goes for hooks.

Edge Bends

12.16. Bring the bend area to heat, localizing this by either pushing the work through the fire or quenching the end before work starts. Put in a set as shown. If the work bows sideways, bring it back to flat with gentle taps.

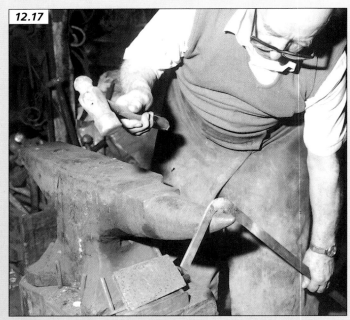

12.17. Form a rough bend over the bick. Again, see how Frank's hammer will fall on the free end: the bick is just a pivot point.

12.18. It's tempting to form the corner over the table, as here. But don't: blows will only draw the metal thinner, necking it down. You want it to bend, not stretch.

12.19. The rough-formed bend has a large radius. To make it tighter (that is, move toward a squarer shoulder) you must push metal toward the corner. Imagine a diagonal line through the required bend. Then, resting the work on the table, strike just off that line. Blows tighten the bend by pushing the top leg down.

12.20. Work alternately on the legs, driving them together and so pushing the joint closer to a right angle. Too much effort will pucker the inside radius, so work gently from both sides. The bend will want to bow sideways so straighten before it goes too far.

12.21. Out comes the try-square (a). The initial bend was too tight, so a couple of light taps brought it back to true (b). See how the anvil is again used as a fixed rest while the hammer does the work? Knowing where to position the work and strike the blow is down to practice. An experienced smith makes it look so easy. It's a bit like the man who charged $160 for five minutes' work in knocking out a dent in a boiler. "$8 pays for my time," he explained to the irate customer. "The rest is for knowing where to hit it."

Punching Holes

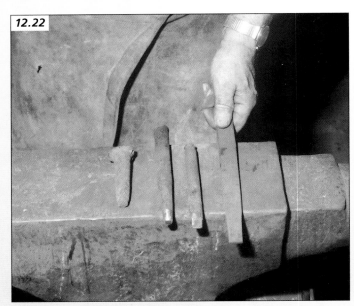

12.22. A selection of punches, used in order of appearance. Right to left: slot punch; flat-ended mandrel or drift; two different-sized round mandrels. All these are easily made to suit different jobs.

12.24. With the corners rounded and someone holding the work, punching begins. Frank uses the anvil table, striking hard to drive the slot punch quickly. After three or four blows, the punch is cooled.

The punch must be centered if metal is to be pushed equally to either side; don't give it a pre-tap before striking in case it moves.

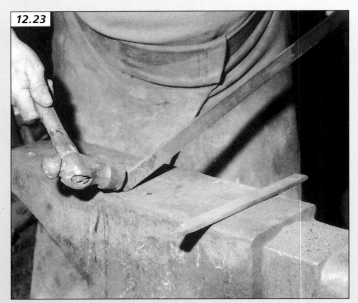

12.23. Frank Dean demonstrates punching technique by putting an eye in the end of flat stock. The same method is used for slots and round holes along the length of a job. For these mark the target area by a row of punch marks, which will remain visible when the work is up to heat. The first operation with an end-eye is to round-up the corners, working the metal between hammer and anvil as shown.

12.25. In goes a pinch of coke dust from a pile on the anvil step. See how much heat is in the work? It's just short of sparkling white. Ease of punching goes up with temperature. On thick work, the punch must be cooled every few blows.

12.26. Frank drives on until feedback through the hammer tells him that the table is nigh. Then he turns the steel and, using the coke dust mark as a guide, drives the slot punch through again. Working over the table hole gives the punch somewhere to go.

12.27. This tiny sliver of metal is all that's lost from the slot. Most has been pushed sideways.

12.28. Now follows the flat-to-round mandrel, again used over the anvil hole. This tool gently takes the oblong to a circle. Bigger, tapered mandrels then follow until the hole is up to size. Very accurate fits can be achieved this way.

12.29. The ring's outer edge is brought true by light blows, using the mandrel and vise as support.

Punching Holes *continued*

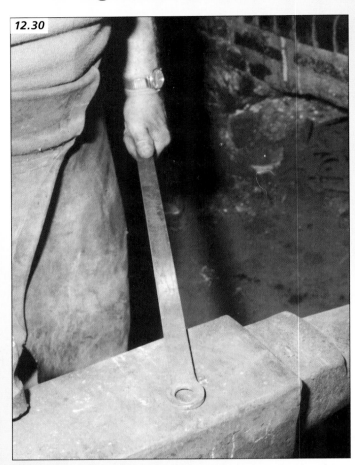

12.30. There's a twist to this ending. With the eye completed, Frank is ready to put a 120° set below it.

12.31. Heat must be localized for the steel to bend only where required. The whole end is brought to heat in the forge, then the target area is isolated by ladling cold water to either side. Gas users can just heat the local area.

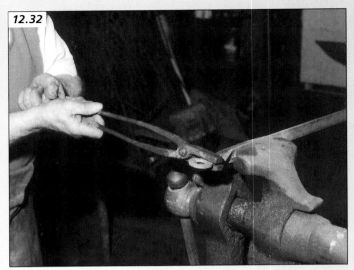

12.32. Into the vise, where a quick twist with the tongs puts a right-angle in the shank . . .

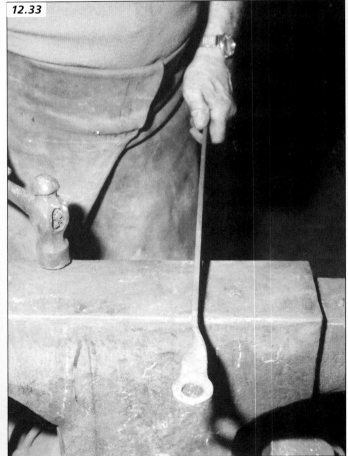

12.33. . . . leaving a neat and tidy ring.

Upsetting

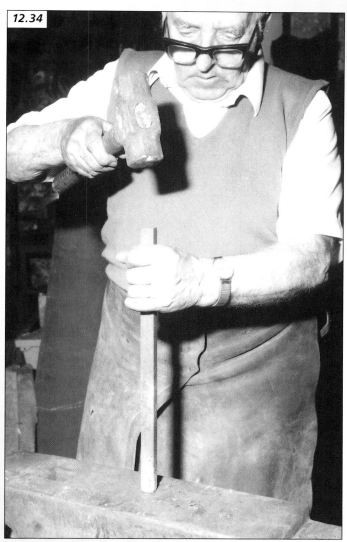

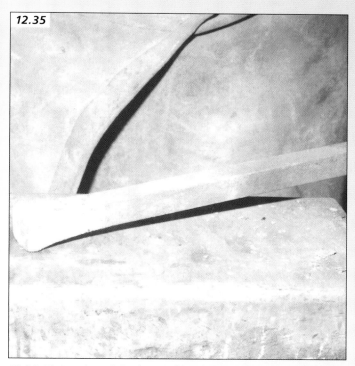

12.35. The end result is plenty of local thickness for subsequent forging operations. Using selective cooling (12.31) the upset area could have been made partway along the bar instead of at the end – an ideal way of concentrating extra metal before punching a large hole.

12.34. Upsetting redistributes metal for later operations. Frank brought the bottom few inches (centimeters) of the bar to a good heat and is now using vertical blows from a big hammer to push the steel up into a bulge. Short work like this can be managed on a normal-height anvil, though the Dean forge has another set flush with the floor for longer stuff. A solid, non-bouncy surface is needed – a concrete floor isn't good enough. The process is repeated, taking fresh heats and re-truing on the way.

Tired of All That Welding?

Every repair job involves making decisions. Not the least of these is allocating blame afterwards or — more importantly — absolving yourself from it. This is often the fitter's most basic and pressing dilemma. The handy cut-out flowchart which follows covers most eventualities. Keep it out of sight as a solution to any given situation. Tested thoroughly in farming, it hasn't failed yet. *(Chart courtesy of the Institute of Diagnostic Engineers)*

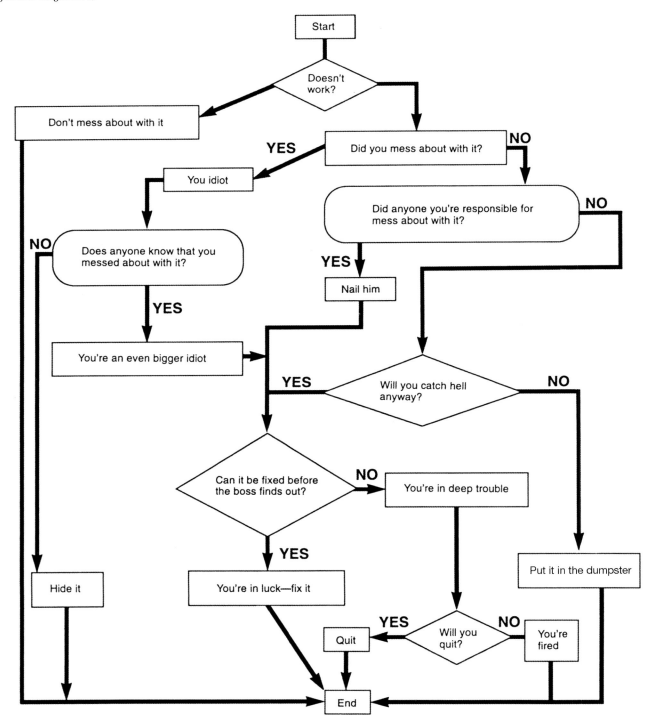

Appendix 1

What's in a Name?

Process Terminology

All technologies have their formal and informal sides. Welding's no exception—depending on who you are and where you are, the same thing goes by different names.

As the formal tags can be hard to get your tongue around, this book sticks with common terms used in the UK. For reference and to help when reading more learned tomes, here's a round-up of the alternatives.

Common name	Formal name
Stick welding, aka MMA (manual metal arc) welding	Shielded metal arc welding (SMAW)
MIG (metal inert gas) welding	Gas metal arc welding (GMAW)
MAG (metal active gas welding)	Gas metal arc welding (GMAW)
Cored wire MIG	Flux-cored arc welding (FCAW)
TIG (tungsten inert gas) welding	Gas tungsten arc welding (GTAW)
Gas or oxyacetylene welding	Oxyfuel welding
Plasma cutting	Plasma arc cutting (PAC)

Appendix 2

Heat Colors

Color	Temp °F (°C)	Use for
Black red (visible only in dull light)	797-824 (425-440)	Pre-heat before welding, toughening carbon steel after quenching
Very dark red	1094-1202 (590-650)	Hardening and annealing carbon tool steel
Dark red	1292-1400 (700-760)	
Cherry red	1499-1697 (815-925)	Hardening alloy tool steel
Orange red	1796-1895 (980-1035)	
Yellow	2003-2192 (1095-1200)	
Yellow white	2300-2399 (1260-1315)	Hardening high-speed steel
White	2498 (1370)	Welding

Note: Temperatures are approximate.

Appendix 3

Useful Conversions

Electrodes and Cored Wire

Need to find the American equivalent of a 2.5mm electrode, or the metric equivalent of 0.045in welding wire? You've come to the right place.

NOTES

Common sizes are shown in bold text. These tables take millimeters as the base unit. The nearest commonly-used imperial equivalents are shown: these may not be an exact conversion.

If you need to make specific conversions, check the internet for an online calculator.

American welding wires are usually specified according to US Steel Wire Gauge. A wire of a given US gauge is generally slightly larger than its equivalent number in British Standard wire gauge (SWG).

SMAW ELECTRODE DIAMETERS

MM	Inch	Wire gauge (SWG)
1.60	1/16	16
2.00	0.08	14
2.50	3/32	12
3.25	1/8	19
4.00	5/32	8
5.00	3/16	6
6.00	1/4	4
8.00	5/16	n/a

Note: With SMAW consumables, 2.4mm electrode is rounded up to 2.5mm

SMAW ELECTRODE LENGTHS

MM	Inch
250	10
300	12
350	14
400	16
450	18
600	24
700	28

TIG ELECTRODES, TIG/GAS FILLER ROD DIAMETERS

With nearest common American inch and wire gauge equivalents

MM	Inch	Decimal inch
0.5		0.020
1.0		0.040
1.6	1/16	0.062
2.4	3/32	0.093
3.2	1/8	0.125
4.0	5/32	0.156
5.0	3/16	0.187

MIG/CORED WIRE DIAMETERS

With nearest common American inch and wire gauge equivalents

MM	Inch	Wire gauge (SWG) [U.K. only]	Decimal inch
0.6			
0.8			
0.9			0.035
1.0			
1.1			0.045
1.2			
1.3			0.052
1.6	1/16	16	
1.7			0.068
2.0	0.08	14	
2.4	3/32	12	

Index

Note: Page numbers in *italics* indicate photo captions, and page numbers ending with a (t) indicate tables. Page numbers in **bold** indicate DIY projects.